Intuitive Interaction

Research and Application

Intuitive Interaction

Research and Application

Edited by
Alethea Blackler

CRC Press
Taylor & Francis Group
Boca Raton London New York

CRC Press is an imprint of the
Taylor & Francis Group, an **informa** business

CRC Press
Taylor & Francis Group
6000 Broken Sound Parkway NW, Suite 300
Boca Raton, FL 33487-2742

© 2019 by Taylor & Francis Group, LLC
CRC Press is an imprint of Taylor & Francis Group, an Informa business

No claim to original U.S. Government works

Printed on acid-free paper

International Standard Book Number-13: 978-1-138-05397-7 (Hardback)

Visit the Taylor & Francis Web site at
http://www.taylorandfrancis.com

and the CRC Press Web site at
http://www.crcpress.com

Dedication

For my little girl, Lizzie.

Contents

PART III   Applying Intuitive Interaction

Preface

Making interfaces more intuitive to use is essential in a world of increasingly complex technologies and increasingly compulsory engagement with these technologies. We need everybody in a society to be able to access banking and government services, entertainment, health care, shopping, and the many other essential and discretionary experiences that technology facilitates. However, many people currently struggle with some of these technologies, and yet more sophisticated interaction experiences are offered to us every day.

The study of intuitive interaction is important because it can help designers, engineers, and usability specialists to create interfaces and products which are actually intuitive, rather than only being marketed as such. Applying the established principles of intuitive interaction to a design process has been shown by several researchers to make an interface easier and more intuitive to use. Obviously, then, applying intuitive interaction can have an impact on the usability of an interface, which should excite designers of all sorts and usability specialists everywhere.

Intuitive interaction is defined as applying existing knowledge in order to use an interface or product easily and quickly, often without consciously realizing exactly where that knowledge came from. Previous research in this field has established this understanding empirically, and relevant past experience has been strongly linked with intuitive interaction in many studies in various contexts across four continents. However, newer work has started to explore a wider range of contexts, applications, and users, and also to consider issues beyond the objective assessment of intuitive interactions.

This volume, the first dedicated to intuitive interaction, explores the most recent research, venturing into new areas not previously discussed in this field, including

- How intuitive interaction can be applied in different contexts where there are many diverse users, such as large-scale urban installations, or a wide range of age groups and how it can apply to the experiences of users in different cultural contexts
- How intuitive interaction is understood and investigated in different academic disciplines (e.g., design and cognitive science) and what implications these differing understandings and approaches may have
- The effect of how users perceive the intuitiveness of an interface and how that may impact their acceptance of technologies and their experience with actual interfaces, and how we can use our understanding of their subjective experience to design more engaging experiences
- Research methods that have been used in intuitive interaction research over the past 18 years and which might be the most effective and most suitable in various situations and contexts
- The commercial applications of intuitive interaction in various contexts and their implications for research and design

Edited by one of the world's foremost researchers in this field, this book includes contributions from authors in academia and industry, with backgrounds in design, psychology, business, IT, and cognitive science, in order to give a fully rounded understanding of the state of the art in intuitive interaction research. This book is designed to equip the reader with the latest knowledge in this important area, along with an understanding of how it can be applied to their own research or study, or to various design, usability, and IT contexts.

Acknowledgments

Many people deserve thanks for helping to get this project started and finished. Dr Dan McAran, author of Chapter 7, started the ball rolling on the book. Executive editor Cindy Carelli and editorial assistant Renee Nakash dealt professionally and efficiently with the editing process at CRC Press from start to finish. All the authors of the various chapters have been insightful to work with and responsive to my many requests. My husband Philip Blackler must have been the world's biggest supporter of intuitive interaction research for the past 18 years. My daughter Elizabeth Blackler has been patient and forbearing with my busy-ness over the past few months. My mother Joanna Wright has provided proofreading, babysitting, cooking, and more over the crucial final few weeks. Dr Claire Brophy was a wonderful assistant during the final editing process. This book would not have happened without you all. My heartfelt thanks.

I would also like to acknowledge the inspiration from my colleagues and PhD students who have worked on various projects in intuitive interaction with me over the years, some of whom have also contributed to this book: Dr Gudur Raghavendra Reddy, Dr Ben Kraal, Dr Simon Lawry, Gowrishankar Mohan, Dr Shital Desai, Professor Doug Mahar, Dr Rafael Gomez, Dr Mitchell McEwan, Dr Claire Brophy, Dr Levi Swann, Professor Daniel Johnson, Associate Professor Peta Wyeth, and Dr Marianella Chamorro-Koc.

Finally, two other intuitive interaction researchers deserve a mention, both names of which will become familiar to readers of this book. My colleague Professor Vesna Popovic has worked with me on this research since 2000 and features in many publications with me (including three in this book). Professor Dr Jörn Hurtienne was one of the first German researchers in this area and was not able to contribute to the book but has been cited extensively by several of the authors, particularly for his work on image schemas. Thanks to you both for your invaluable contributions to intuitive interaction research.

Editor

Associate Professor Alethea Blackler (PhD) is associate director of the QUT Design Lab at Queensland University of Technology, Brisbane, Australia, and was formerly head of discipline for industrial design at the university. Her principal area of research interest for the past 18 years has been intuitive interaction, in which she is one of the world leaders. She pioneered the first empirical work in the field, has led various funded and commercial projects on intuitive interaction, and has edited a special issue of the journal *Interacting with Computers* on intuitive interaction in 2015. She also works in the area of older people and technology, and her latest project is editing this volume on intuitive interaction as well as authoring and co-authoring four of the chapters. She has attracted and worked with external partners in government, community, and industry on various projects, including prestigious Australian and European competitive grants. She is a very experienced research degree supervisor, with 7 current PhD students and 11 PhD completions. She has published more than 60 peer-reviewed papers, has been invited to give presentations internationally, and is the recipient of several awards (a.blackler@qut.edu.au).

Contributors

Alethea Blackler
QUT Design Lab
Queensland University of Technology
Brisbane, Australia

Shital Desai
QUT Design Lab
Queensland University of Technology
Brisbane, Australia

Sarah Diefenbach
Ludwig Maximillian University of
    Munich
Department of Psychology
Munich, Germany

Sandrine Fischer
Independent Researcher
San Fransisco, California

Christopher R. Hammond
IBM
Austin, Texas

Luke Hespanhol
School of Architecture, Design and
    Planning
University of Sydney
Sydney, Australia

Dan McAran
Henley Business School
University of Reading
Reading, United Kingdom

Mitchell McEwan
Games Research and Interaction
    Design Lab
Queensland University of Technology
Brisbane, Australia

Marita A. O'Brien
Franciscan University of Steubenville
Department of Psychology
Steubenville, Ohio

Temitope A. Ogunyoku
IBM Research Africa
Catholic University of Eastern Africa
Nairobi, Kenya

Jesyka M. Palmer
IBM
Austin, Texas

Vesna Popovic
QUT Design Lab
Queensland University of Technology
Brisbane, Australia

Gudur Raghavendra Reddy
Faculty of Arts and Design
University of Canberra
Canberra, Australia

Jeremiah D. Still
Department of Psychology
Old Dominion University
Norfolk, Virginia

Mary L. Still
Department of Psychology
Old Dominion University
Norfolk, Virginia

Stefan Tretter
Ludwig Maximillian University of
    Munich
Department of Psychology
Munich, Germany

Daniel Ullrich
Ludwig Maximillian University of
    Munich
Institute of Informatics
Munich, Germany

Part I

*Intuitive Interaction
Theory and Context*

1 Intuitive Interaction
An Overview

Alethea Blackler

CONTENTS

1.1 INTRODUCTION

Making interfaces, systems, products, and even environments more intuitive to use is essential in a world of increasingly complex and ubiquitous technologies. Everybody in a society needs to be able to easily access a wide range of services provided in diverse ways. However, currently, people with lower technology familiarity (e.g., older people and those facing various types of disadvantage) struggle with some of these technologies. Even middle-aged people show a decrease in familiarity when they are faced with interfaces they do not own (Lawry, Popovic, and Blackler, 2011). And yet more sophisticated technologies and interactive experiences are offered to us every day. Researchers in intuitive interaction believe that they can help designers to understand how to design interfaces and experiences that can be truly intuitive for all sorts of people to use. This book showcases much of the state-of-the-art work in this field, so that the reader has a full picture of how we have come to define and understand intuitive interaction through research and how it can be best applied to a great variety of products, interfaces, environments, and more.

Intuitive interaction is defined as interaction with an interface that is informed by past experience. Normally it is fast, generally accurate, and often somewhat non-conscious (Blackler, Popovic, and Mahar, 2003). This book explores some of the most recent research in intuitive interaction, building on past work and venturing into new areas such as the effects of how users perceive the intuitiveness of an inter-face, how people experience intuitive interaction subjectively, and how we can use our understanding of intuitive interaction to design less effortful and more engaging experiences. It looks at how intuitive interaction can be applied in different contexts, such as large-scale urban installations, the developing world, or in older popula-tions, and in different industries. The book also addresses how intuitive interaction is understood in different academic disciplines (e.g., design and cognitive science) and what methods have been used in its research over the past 18 years. This chap-ter introduces the concept of intuitive interaction, covers a brief history of intui-tive interaction research, and then provides an overview of the later chapters before detailing the important contributions made by the book and the implications these will have for future research in intuitive interaction.

1.2 HISTORY OF INTUITIVE INTERACTION RESEARCH

Although the concept was much talked about during the latter part of the last cen-tury, no empirical research was done into intuitive interaction until 2000. Two dis-parate groups worked on it separately before making contact in 2006. These were Blackler and colleagues in Australia and the Intuitive Use of User Interfaces (IUUI) research group in Germany. Between them, these groups conducted most of the early work, but since then other researchers from around the world have expanded and built on their contributions.

1.2.1 EARLY EMPIRICAL WORK

Blackler, Popovic, and Mahar (2010) conducted three lab-based experiments, with a total of 110 participants. They used real, contemporary products as media-tors to reveal participant knowledge: an early digital camera and a reconfigurable, touchscreen universal remote control. Technology familiarity (TF) was used in all three experiments to gauge past experience with relevant interface features. It was measured through a questionnaire in which participants provided details of their experience with products and interfaces with similar features to those they would encounter during the experiment (Blackler, Popovic, and Mahar, 2010). This TF tool has been used, adapted, and applied in various contexts in the years since (Brophy, Blackler, and Popovic, 2015; Cave, Blackler, Popovic, and Kraal, 2014; McEwan, Blackler, Johnson, and Wyeth, 2014; O'Brien, Weger, DeFour, and Reeves, 2011; Still and Still, 2018).

In each experiment, participants were video recorded performing set tasks with the products while delivering concurrent (think-aloud) protocol. Then they were interviewed about their familiarity with each interface feature they used. Audiovisual data were coded for correctness and the intuitiveness of feature interac-tion using literature-based coding heuristics (Blackler, Mahar, and Popovic, 2010;

Blackler, Popovic, and Desai, 2018). All three experiments showed that intuitive interaction is based on past experience with similar products and interface features. Familiar features were used more intuitively, and people with higher TF completed tasks more quickly, with more intuitive uses and fewer errors. Results also suggested that older people, even those with higher TF, completed tasks more slowly and less intuitively than younger people. The early empirical work helped to establish a definition of intuitive interaction as based on past experience. This was quickly backed up by other research and TF or equivalent measures in each context have consistently emerged as the most influential predictors of intuitive use in many studies in various contexts across four continents (Blackler, Mahar, and Popovic, 2010; Fischer, Itoh, and Inagaki, 2015a; Hurtienne and Blessing, 2007; Hurtienne and Israel, 2007; Mohs et al., 2006; O'Brien, Rogers, and Fisk, 2008).

1.2.2 IMAGE SCHEMAS

In Germany, one of the major approaches to designing for intuitive interaction has been through applying *image schemas*. Hurtienne (2009), during his involvement in the IUUI group in Germany, conducted a range of studies examining the role of image schemas in intuitive use, drawing on Lakoff and Johnson's (1980) work. Image schemas are "abstract representations of recurring dynamic patterns of bodily interactions that structure the way we understand the world" (Hurtienne and Israel, 2007, p. 130) and are thus important building blocks for thinking. They are based on each individual's experience of interaction with the physical world but tend to be largely universal, as the physical world operates in the same way for almost everyone who is able-bodied. Because they are based on past experience, and because they are so well known and so universal that they become unconscious, image schemas can be defined as intuitive. Therefore, Hurtienne argued, incorporating image schemas into interfaces can allow intuitive interaction. For example, the up–down image schema is established by experience of verticality and is applied to our understanding of a range of other concepts, such as quantity. *Up* is associated with *more* of a substance because, for example, a glass is fuller when the substance within it reaches the top or piles of objects are higher when in greater number.

Hurtienne used custom software to test the metaphorical extensions of image schemas applied to interface design in four experiments using the up–down and near–far image schemas. All these studies supported the notion that using metaphorical extensions of image schemas in the software contributed to accurate and intuitive use. Hurtienne's final three studies used a participant-generated and verified list of real-world examples of image schemas applied to redesigning commercially available accounting software. Users of the original software evaluated the new interface designs. The results showed that both the redesigned graphical user interface and a hybrid graphical/tangible interface were rated higher than the existing system.

Through his research, Hurtienne (2009) demonstrated that metaphorical extensions of image schemas can be used in interface design and result in better performance. The effective use of image schemas and their metaphorical extensions is likely to facilitate intuitive use, as image schemas are based on largely unconscious prior knowledge that almost every person possesses. Thus, performance using interfaces based on image schemas should remain consistent across heterogeneous user groups, making them more ubiquitously (or universally) applicable than familiar interface features, which may not be familiar to everyone and generally rely on experience with similar and dissimilar products. Hurtienne (2009) identified around 40 image schemas based on human experiences that can facilitate intuitive interaction. The image schema method is relatively simple and easy to implement in design; however, the interpretation of metaphorical extensions can be culturally sensitive.

1.3 EXPANSION OF INTUITIVE INTERACTION RESEARCH DOMAINS

Subsequent research on intuitive interaction expanded the field by building on the established theory and measurement tools through their application in a range of domains, such as website design (Mohan, Blackler, and Popovic, 2015), public spaces (Hespanhol and Tomitsch, 2015), video games (McEwan et al., 2014), and the study of passenger navigation of airport terminals (Cave et al., 2014). However, these later applications have mainly focused on three major themes: the relationship between aging, familiarity, and intuitive use, primarily drawing on Blackler and colleagues' work (Blackler, Popovic, Mahar, Reddy, and Lawry, 2012; Gudur, 2012; Mihajlov, Law, and Springett, 2014; O'Brien, 2010); the use of image schemas in the design process to encourage intuitive use, following up on the IUUI work (Chattopadhyay and Bolchini, 2015; Fischer et al., 2015a; Hurtienne, Klöckner, Diefenbach, Nass, and Maier, 2015); and the relationship between so-called natural user interfaces, such as touch, tangible, and embodied interfaces, and intuitive interaction (Antle, Corness, and Droumeva, 2009; Chattopadhyay and Bolchini, 2015; Desai, Blackler, and Popovic, 2015, 2016; Macaranas, 2013; Mihajlov, Law, and Springett, 2014). This section will discuss these three areas in more depth.

1.3.1 AGE

Much of the work following up on Blackler and colleagues' initial explication of the relationship between technology familiarity and intuitive use examines the additional influencing role of age-related factors on intuitive interaction (Blackler et al., 2012; Mihajlov et al., 2014; O'Brien, 2010; Gudur, 2012). O'Brien (2010) used a technology experience questionnaire (based on Blackler's TF questionnaire) to group older (over 65) and younger (under 45) adults by their TF level. She found that older adults used different categories of technologies, including more health-care and kitchen products, than younger adults. She also showed that prior experience was the most common reason for successful technology use but was not always sufficient on its own. Information presented by the products themselves was also needed to address problems encountered in real-life tasks. Both low- and high-TF older adults using a video camera, digital radio alarm clock, and e-reader did not perform as well as younger adults.

The Australian program of research (Blackler et al., 2012) meanwhile included five new empirical studies focused on intuitive interaction for older people, using continuous age groups from 18 to 83. Findings showed that older people were less familiar with products that they owned than younger people and used fewer products overall, while both older and middle-aged people were less familiar with products that they did not own than younger people (Lawry, Popovic, and Blackler, 2010; Lawry et al., 2011). Older people were significantly slower at completing tasks and showed significantly fewer intuitive uses and correct uses. However, with an interface designed using a tool for designing for intuitive interaction, they performed better, had significantly more intuitive uses, and rated the interface as significantly more familiar than a comparable interface (Blackler, Popovic, and Mahar, 2014). Age-related cognitive decline is also related to slower and less intuitive performance with contemporary products and interfaces (Blackler, Mahar, and Popovic, 2010; Reddy, Blackler, Mahar, and Popovic, 2010). Therefore, the reasons behind the problems that older people demonstrate with contemporary technologies involve a mix of familiarity and capability. Redundancy applied to an interface in the form of symbols and words is helpful for middle-aged and younger old people, but the oldest age group performed better with a words-only interface, suggesting that clutter in the redundant interface could have been a factor for the oldest group (Gudur, Blackler, Popovic, and Mahar, 2009). Also, older people showed faster and more intuitive use with a flat interface than a nested one, although there was no difference in errors (Gudur, Blackler, Popovic, and Mahar, 2013).

Further research with technology-naïve older adults has confirmed some of these findings around interface complexity, noting that complex (multiple-finger) touch interactions are less intuitive than simple drag gestures for older users (Mihajlov et al., 2015). McEwan's (2017) results also broadly suggested that older users have less TF, which results in lower intuitive use, yet showed that interfaces with higher levels of natural mapping could provide some compensatory effect for this reduced familiarity (presumably by leveraging sensorimotor knowledge not captured in TF measurements). Hurtienne et al. (2015) reported on a study where they applied image schemas in order to improve the intuitiveness of an interface. Subsequent testing showed that subjective measures of intuitive use were positive, with no significant differences between ages.

The research on intuitive interaction and aging has shown that age does impact the intuitive use of a variety of interfaces and that this is due to a complex mix of familiarity and cognitive decline (along with the already well-established implications of physical and sensory decline). Designing appropriately for them can increase the intuitiveness of interfaces for older people but does not always work as expected so needs to be applied and tested carefully. This is discussed further in Chapters 5 (O'Brien, 2018) and 6 (Reddy, Blackler, and Popovic, 2018), where experiments with older people and a variety of interfaces are described and recommendations are provided, as well as a proposed framework for design intuitive adaptive interfaces to cater for people with a more diverse range of TF (such as older populations).

1.3.2 FURTHER WORK WITH IMAGE SCHEMAS

Following up on the IUUI work, a number of researchers have examined the role image schemas can play when designing interfaces to offer a greater intuitive use potential. In Japan, Fischer et al. (2015a) conducted an empirical experiment to study the effects of schema induction, feature familiarity, goal familiarity, and cognitive style on performance using a vehicle interface. They studied 31 young adults navigating through an interface containing approximately 160 features whose functions ranged from ranged from basic (e.g., dialing a number, defrosting the windscreen) to speculative (e.g., wirelessly retrieving advertisements). The features were grouped into 75 states that were in turn grouped into five technology domains: audio, air conditioning, navigation, onboard computers, and telephones. The results revealed that the induction of a new schema improved the exploration of states for new features but not for familiar features. The knowledge required to use the features intuitively is derived not only from across devices but also from across domains, and familiar features were processed through the transfer of prior schemas.

Hornecker (2012) and other HCI researchers have critiqued the intuitive interaction design principles as relying too heavily on the transfer of knowledge from existing interfaces, which they say limits the novelty of design solutions. However, in the project mentioned above, Hurtienne et al. (2015) applied image schemas to an interface during the design process in order to see whether it is feasible to achieve inclusiveness, intuitive use, and innovation in one product. Subjective measures of both innovation and intuitive use were in the positive half of their respective scales, so the results suggested that the careful application of intuitive interaction can achieve innovative interfaces.

Other researchers have also aimed to evaluate the intuitive interaction benefits of image schemas across a range of interactive systems. Fischer et al. (2015b) proposed a new, more automated means to evaluate the knowledge transfer of relevant image schemas to guide the redesign of proposed interface prototypes. In their work, the existing methods for measuring intuitive interaction were criticized as being time intensive for designers/researchers and users alike as well as focusing too much on the symptoms of intuitive use rather than quantitatively evaluating the transfer of previous knowledge. They argued that design strategies aimed at improving intuitive use are most successful in novel/innovative or complex systems and that applying these strategies to systems where the interface is simple (i.e., the image schemas are basic and clear) or where users are highly experienced may offer little gain in intuitive potential.

Asikhia, Setchi, Hicks, and Walters (2015) also worked on the evaluation of interfaces, eliciting image schemas from existing interfaces and using them during usability testing to test interfaces for their intuitiveness. Approaches to user testing for intuitive interaction have so far been based on the same methods as experiments for investigating it, making them time-consuming and cumbersome in the usability-testing context (Blackler, Popovic, and Desai, 2018; Blackler et al., 2011), so these approaches are very promising. Image schemas continue to offer a potential avenue for intuitive use in a variety of new contexts and user groups, as well as offering an avenue to designing both innovative and intuitive interfaces. How the application of image schemas relates to other theories around intuitive interaction is discussed in more detail in Chapter 2 (Blackler, Desai, McEwan, Popovic, and Diefenbach, 2018).

1.3.3 SUMMARY

As can be seen in this brief overview, intuitive interaction research has been progressing over the past 18 years. This book is intended to showcase the latest research in this area as well as give examples of the application of intuitive interaction theories and recommendations into real-world domains. The rest of Chapter 1 will discuss the chapters in the book, exploring their content and contributions to the understanding and application of intuitive interaction.

1.4   BOOK CHAPTERS

This book is structured into three sections. Part I, "Intuitive Interaction Theory and Context," covers Chapters 1 through 3. It sets the scene with a history of the field and provides a theoretical framework and research-based theory for the following chapters. Part II, "Research and Findings in Intuitive Interaction," includes Chapters 4 through 8. It covers research methods used for intuitive interaction research and showcases a variety of recent findings in intuitive interaction across a diverse range of contexts. Part III, "Applying Intuitive Interaction," contains Chapters 9 through 11, which include case studies of intuitive interaction theory and recommendations applied to actual products and interfaces in various contexts in the real world.

1.4.1   Part I: Intuitive Interaction Theory and Context

Chapter 1, "Intuitive Interaction: An Overview," presents a short history of intuitive interaction research, from the early empirical work on establishing a definition and applications such as images schemas through to work on intuitive interaction for older people. It will introduce the rest of the chapters in the book and conclude by discussing their contributions and implications as well as directions for future research in intuitive interaction.

Chapter 2, "Perspectives on the Nature of Intuitive Interaction," discusses theoretical ideas and recommendations previously developed in intuitive interaction research as well as exploring newer issues in the field, such as the subjective experience of intuitive interaction and the potential of *tangible and embodied interfaces* (TEIs) and *natural user interfaces* (NUIs) to be intuitive. It then presents the Enhanced Framework for Intuitive Interaction (EFII) (Figure 2.2), aimed at assisting readers in understanding and applying intuitive interaction, based on empirical research conducted around the world over 18 years. EFII is included in order to give the reader a firm understanding of the theory developed in intuitive interaction research and how each concept relates to others, and will help readers to understand and apply the concepts presented throughout the rest of the book.

Chapter 3, "Cognitively Describing Intuitive Interactions," discusses the cognitive bases of intuitive interaction so that the reader can understand how and why it happens and demonstrates how interventions to improve the intuitiveness of interfaces may work. In this chapter, the authors have identified cognitive processes that play a central role in supporting intuitive interaction, considering how knowledge representations might constrain the methods used to effectively elicit that knowledge. To convey these cognitive principles and their relation to design, the authors present the Intuitive Interaction Hierarchy (Figure 3.1), which helps to position intuitive interaction research alongside cognitive science concepts. The chapter goes on to use this understanding to explain an experiment on various instruments used to measure intuitiveness and discusses the implications of the findings for researchers and practitioners.

1.4.2   PART II: RESEARCH AND FINDINGS IN INTUITIVE INTERACTION

Chapter 4, "Research Methods for Intuitive Interaction," presents the various methods that have been used in intuitive interaction research. It discusses in depth the core method used in this research: the observation of participants using interfaces. It goes on to discuss measures of intuitive interaction and familiarity, both objective and subjective, and their pros and cons. Finally, it describes and compares the various data collection methods used in this field over the years. Many involve standard instruments that have been adapted, but others have included the production of unique measures and coding schemes. The methods are presented as the Intuitive Interaction Research and Application Toolkit (Table 4.1) and are discussed and explained so that readers will be able to apply them to their own research and practice.

Chapter 5, "Lessons on Intuitive Usage from Everyday Technology Interactions among Younger and Older People," discusses a series of studies with older people, both longitudinal and observational. Both intuitive and reflective interaction processes were identified through the analysis of observations, and the overall finding was that participants' technology experience was not the only factor determining the success of technology encounters but that a more diverse description of the relevant knowledge was also needed. Low technology experience did not prevent several older people from successfully completing tasks, and many high-tech older people seemed to use the same reflective approach to completing tasks as low-tech older people when they noticed mistakes in their interaction. Finally, an in-depth discussion of the implications of the findings for designing for older people is included. The importance of feedback and feedforward from devices is particularly highlighted as essential for older people to be able to find their way through a task. Feedforward could prevent instruction-led first use and build user knowledge and confidence for exploring new features. Feedback that enables rapid problem detection and recovery can help users feel confident performing basic operations and exploring additional features. Overall, the findings suggest opportunities to improve technology usage through the appropriate design of devices.

Chapter 6, "Adaptable Interface Framework for Intuitively Learnable Product Interfaces for People with Diverse Capabilities," further addresses the complex issues involved in designing for older people and presents a framework called the Intuitively Learnable Adaptable Interface System (ILAIS) (Figure 6.1). Previous research conducted by the authors found, just like O'Brien (2018), that older people often offset speed for accuracy and that they are able to use interfaces correctly if they are designed appropriately and plenty of time is available. The pros and cons of various types of adaptive and adaptable interfaces are explained and discussed. ILAIS is intended for developing adaptable interfaces that have the potential to be intuitively learnable for a range of users, in particular older people, who have lower technology familiarity. This means they are simple to learn and will become intuitive over time, as it may not be possible to make all interfaces intuitive for all people at first-time use. ILAIS is based on both the literature and experiments into interventions intended to make interfaces more intuitive for older people. It is justified and exemplified in the chapter so that it can be applied by other researchers and practitioners.

Chapter 7, "Development of the Technology Acceptance Intuitive Interaction Model," presents the first work done on intuitive interaction and the well-known Technology Acceptance Model (TAM). It presents empirical work that suggests *perceived intuitiveness* is a viable new variable as part of the TAM and that the concept of intuitiveness may be replacing the concept of *perceived ease of use* (PEOU). A new version of the TAM, the Technology Acceptance Intuitive Interaction (TAII) model (Figure 7.3), is described, and the testing of its viability is detailed. This work complements the work in Chapter 8 in exploring how the perceived intuitiveness of an interface affects how people accept and interact with it.

Chapter 8, "Intuitive Interaction from an Experiential Perspective: The Intuitivity Illusion and Other Phenomena," introduces the concept of the *intuitivity illusion*. It presents research that suggests that how people perceive and experience an interface as intuitive may be just as, if not more, important as how intuitive researchers or designers may assess it to be. The authors present the INTUI model (Figure 8.1), which helps to frame how this illusion was understood and discovered through empirical work. The implications of this work are important, as technology adoption may be related to this illusion more than it is related to actual intuitive interaction. Therefore, it needs to be understood by designers and marketers. This relates closely to the work presented in Chapter 7, as it suggests that how people perceive the intuitiveness of an interface has an important impact on whether they will accept or adopt it. Both Chapters 7 and 8 suggest that there is more work to be done in the area of perceived intuitiveness and on understanding exactly how it relates to quantifiable intuitiveness, which leads to more successful interface use. On top of all these specific theoretical implications, the intuitivity illusion implies the need for a growing focus on instructions (or other sources of observational learning) when dealing with intuitive interaction.

1.4.3 PART III: APPLYING INTUITIVE INTERACTION

Chapter 9, "City Context, Digital Content, and the Design of Intuitive Urban Interfaces," looks at interactive urban interfaces and how they relate to the concept of intuitive interaction, which is an important approach, as these types of interfaces are intended to be used by a very wide range of people with instant but only occasional access. It presents a typology, the Urban Interaction Archetypes (Urbia) model (Figure 9.2), which explains how interfaces of this type can be understood in the context of intuitive interaction, explaining this through examples of applications of each archetype. The chapter then presents three real-world examples of the misappropriation of these archetypes by members of the public and discusses the implications of this when designing urban interfaces for intuitive interaction.

Chapter 10, "Designing Intuitive Products in an Agile World," discusses the reality of applying intuitive interaction to real commercial projects using lean and agile methods for identifying and optimizing those features and functions of an interface most likely to be used intuitively by target user groups. It applies theory to explain ways in which features can be selected for aesthetic treatment or focused design or redesign. It introduces and explains the Intuitive Product Development (IPD) canvas (Figure 10.2), the first effort to align the development of new features in terms of users' prior schemas, and then presents two cautionary principles that help to guide designers toward the most efficient process for developing intuitive interfaces. It then presents a case study of a *sprint* that applied the canvas and principles in a real-world project. This shows how these agile methods can be used in industry to develop interfaces within short time frames but retaining rigorous approaches and therefore achieving reliable results from user feedback.

Chapter 11, "Intuitive Interaction in Industry User Research: Context is Everything," is a case study that discusses an IBM project undertaken in Africa to design a public access health kiosk that was user tested by both local and international participants. It reveals cross-cultural issues not previously explored in the context of intuitive interaction, especially around how designers can understand and apply relevant technology familiarity and expectations, which may be very culturally dependent, and cultural differences in the ways in which people provide subjective feedback about interfaces.

1.5 CONTRIBUTIONS OF THIS BOOK

This book has contributed much to the field of intuitive interaction research. Firstly, as presented in Chapter 1 (Blackler, 2018), Chapter 2 (Blackler, Desai, et al., 2018), and Chapter 3 (Still and Still, 2018), there is now an increased understanding of intuitive interaction and how it works, and how the various ideas and theories in the field relate to each other. Secondly, useful methods for intuitive interaction research are presented throughout the book (Chapter 3 [Still and Still, 2018], Chapter 5 [O'Brien, 2018], Chapter 7 [McAran, 2018], Chapter 8 [Tretter, Diefenbach, and Ullrich, 2018], and Chapter 10 [Fischer, 2018]), and all of these and more are explored in depth and compared and presented in a concise toolkit in Chapter 4 (Blackler, Popovic, and Desai, 2018). Thirdly, there are findings in various areas presented in the different chapters; for example, findings about designing more intuitive interfaces for older people and discussions of a range of interventions that could work for them (Chapters 5 [O'Brien, 2018] and 6 [Reddy et al., 2018]). Fourthly, there are several frameworks to assist in understanding and applying intuitive interaction presented in various chapters (detailed in Section 1.5.1), and Part III (Chapters 9 through 11) also contains case studies that practitioners can learn from when applying intuitive interaction to their designs.

1.5.1 MODELS, FRAMEWORKS, AND TOOLKITS

Chapter 2 (Blackler, Desai, et al., 2018) presents EFII, intended to be an overarching representation of how all the ideas in this field relate to each other and how they can work together to help designers create more intuitive interfaces. Chapter 3 (Still and Still, 2018) contains the Intuitive Interaction Hierarchy, which specifically aims at helping to understand how cognitive science principles relate to the ideas in intuitive interaction research. Chapter 4 (Blackler, Popovic, and Desai, 2018) presents the Toolkit of Intuitive Interaction Research Methods, which summarizes the various methods suitable for use in different contexts, depending on the goals and objectives of the research. Chapter 6 (Reddy et al., 2018) presents ILAIS, which is aimed at developing adaptive interfaces that can become intuitive by helping users to gain and apply new knowledge as they use the interface. Chapter 7 (McAran, 2018) contains the TAII model, a significant step in bringing understanding of intuitive interaction into the world of technology acceptance research. Chapter 8 (Tretter et al., 2018) summarizes the INTUI model, which has been presented previously but here is applied to understanding the important phenomenon of the intuitivity illusion. Chapter 9 (Hespanhol, 2018) unveils the Urbia model for understanding and applying intuitive interaction in interactive urban interfaces or large-scale interactive installations. Chapter 10 (Fischer, 2018) presents the IPD canvas, which allows practitioners and developers to apply intuitive interaction to their projects in an efficient yet rigorous way. These models, frameworks, and toolkits are all useful for researchers and designers in aiding understanding and facilitating the application of intuitive interaction. All of them are complementary as all are based on previous empirical work and current research findings, reflecting and building on the established understanding of intuitive interaction discussed in Sections 1.2 and 1.3 and presented in EFII (Blackler, Desai, et al., 2018). This means that readers can apply whatever framework or model is appropriate to their situation without conflict or confusion.

1.5.2 Emerging Concepts and Future Work in Intuitive Interaction

Some important and common concepts emerge from several of the chapters in the book that deserve to be highlighted. These offer new ideas or synergies in intuitive interaction research and present opportunities for future work in the field.

Feedback is highlighted by O'Brien (Chapter 5, 2018) and Hespanhol (Chapter 9, 2018) as an important component of intuitive interaction design. O'Brien (Chapter 5, 2018) discusses feedback and feedforward and Hespanhol (Chapter 9, 2018) mentions the specificity, timing, and modality of feedback. Giving appropriate and timely feedback is an established method of allowing forgivable and easy-to-use designs and interfaces but has not previously been discussed in depth in this research community. It is now worth exploring how it can be better incorporated into recommendations for design for intuitive interaction.

Perceived intuitiveness is a concept that has recently emerged and is discussed in Chapter 7 (McAran, 2018) and Chapter 8 (Tretter et al., 2018). Thanks to these contributions, we can begin to understand the implications of the ways in which people perceive the intuitiveness of interfaces and how this affects design and marketing, as well as ways to measure the subjective experiences of intuitiveness. Tretter et al. (Chapter 8, 2018) suggest that one good way to address this is to prime people by providing marketing demonstrations and material that show users how to do certain operations before they even use an interface, suggesting that instructions themselves can be considered a *design element*, especially in complex IT products where they can now be built in. This sort of idea (a simple brochure) was applied by Fischer (Chapter 10, Fischer, 2018) as part of prototype testing and had a positive effect. This concept also has interesting implications for the design of adaptive interfaces (Chapter 6, Reddy et al., 2018), which could potentially include instructions or *gamified* practice routines as well as assessment exercises, all of which can be used to form adaptation rules and algorithms.

Chapter 10 (Fischer, 2018) highlights the issue of innovation versus intuitive interaction, which is also mentioned in Chapter 2 (Blackler, Desai, et al., 2018), and goes some way toward showcasing ways in which intuitive interaction and innovation can be combined in interfaces by applying Fischer's *cautionary principles*.

The subjective experience of intuitive interaction is becoming more of a focus for research in this field (e.g., Chapter 2 [Blackler, Desai, et al., 2018], Chapter 8 [Tretter et al., 2018], and Chapter 7 [McAran, 2018]). There is more work to be done in this area so that we can learn how to combine fast and effortless use with engaging subjective experiences to make better interfaces for everyone. Manipulating transfer distance could be one way to achieve this.

We could explore the possibility of designing things specifically so that they become intuitive through practice or instruction (Chapter 8 [Tretter et al., 2018], Chapter 10 [Fischer, 2018]), or through the use of various levels of adaptive interfaces (Chapter 6 [Reddy et al., 2018]). This has been called *inferential intuition* (Mohs et al., 2006) but has not been thoroughly explored to date and is worthy of more attention. This idea also has implications for the definition of intuitive interaction itself and the question of whether it should be expanded to include the idea of training/instruction and the implications of an interface becoming intuitive over time, whether through repeated interaction or design.

Culture and intuitive interaction has been little explored in the past, but several issues relating to this area are raised in this book: for example, the issues with NUIs and culturally specific gestures raised in Chapter 2 (Blackler, Desai, et al., 2018), the US-specific example of steering signals given in Chapter 3 (Still and Still, 2018), and the culturally mediated issues with both prior knowledge and subjective feedback revealed in Chapter 11 (Palmer, Ogunyoka, and Hammond, 2018). Since intuitive interaction is so closely related to past experience, we need to start to consider more precisely which experiences are culturally based and which could be universal. The *pathways to intuitive use* presented in EFII (Chapter 2; Blackler, Desai, et al., 2018) could be a good starting point for understanding this as they already include categories that separate universal knowledge, such as physical affordances, from cultural knowledge, such as population stereotypes. However, as Palmer et al. (2018) caution in Chapter 11, practitioners need to better understand what their precise target audience knows and expects in order to design intuitive interfaces for them.

1.6 CONCLUSION

This chapter has presented an overview of previous intuitive interaction research and an overview of the contents of this book, which includes much of the current research in the field. The book contains contributions from authors with a diverse range of backgrounds (design, IT, psychology, business, cognitive science, and usability) and is thoroughly grounded in empirical research from around the world but also offers practical examples of applying these ideas in various industries on different continents. The highlighted contributions show that there is much for the field to gain from this book and the future work suggests signposts where new research might head. I hope that readers find the information interesting and valuable and most of all that designers of all types are able to apply these ideas so that users can enjoy more intuitive experiences with the objects, interfaces, and services that surround them in ever-increasing layers of complexity and ubiquity.

REFERENCES

Antle, A. N., Corness, G., and Droumeva, M. (2009). Human–computer intuition? Exploring the cognitive basis for intuition in embodied interaction. *International Journal of Arts and Technology*, 2(3), 235–254.

Asikhia, O. K., Setchi, R., Hicks, Y., and Walters, A. (2015). Conceptual framework for evaluating intuitive interaction based on image schemas. *Interacting with Computers*, 27(3), 287–310.

Blackler, A. (2018). Intuitive interaction: An overview. In A. Blackler (Ed.), *Intuitive Interaction: Research and Application* (pp. 3–18). Boca Raton, FL: CRC Press.

Blackler, A., Desai, S., McEwan, M., Popovic, V., and Diefenbach, S. (2018). Perspectives on the nature of intuitive interaction. In A. Blackler (Ed.), *Intuitive Interaction: Research and Application* (pp. 19–40). Boca Raton, FL: CRC Press.

Blackler, A., Mahar, D., and Popovic, V. (2010). Older adults, interface experience and cognitive decline. Paper presented at OZCHI 2010, "Design—Interaction—Participation," 22nd Annual Conference on the Australian Computer–Human Interaction Special Interest Group, Brisbane, Australia.

Blackler, A., Popovic, V., and Desai, S. (2018). Research methods for Intuitive Interaction. In A. Blackler (Ed.), *Intuitive Interaction: Research and Application* (pp. 65–88). Boca Raton, FL: CRC Press.

Blackler, A., Popovic, V., Lawry, S., Reddy, R. G., Doug Mahar, Kraal, B., and Chamorro-Koc, M. (2011). Researching intuitive interaction. Paper presented at IASDR2011, 4th World Conference on Design Research, Delft, the Netherlands.

Blackler, A., Popovic, V., and Mahar, D. (2003). The nature of intuitive use of products: An experimental approach. *Design Studies*, 24(6), 491–506.

Blackler, A., Popovic, V., and Mahar, D. (2010). Investigating users' intuitive interaction with complex artefacts. *Applied Ergonomics*, 41(1), 72–92.

Blackler, A., Popovic, V., and Mahar, D. (2014). Applying and testing design for intuitive interaction. *International Journal of Design Sciences and Technology*, 20(1), 7–26.

Blackler, A., Popovic, V., Mahar, D., Reddy, R. G., and Lawry, S. (2012). Intuitive interaction and older people. Paper presented at DRS 2012, "Research: Uncertainty, Contradiction and Value," Bangkok, Thailand.

Brophy, C., Blackler, A., and Popovic, V. I. E. (2015). *Aging and everyday technology*. Proceedings of the 6th International Association of Societies of Design Research (IASDR) Congress, Brisbane, Australia.

Cave, A., Blackler, A. L., Popovic, V., and Kraal, B. J. (2014). Examining intuitive navigation in airports. Paper presented at DRS 2014, "Design Big Debates: Pushing the Boundaries of Design Research," Umea, Sweden.

Chattopadhyay, D., and Bolchini, D. (2015). Motor-intuitive interactions based on image schemas: Aligning touchless interaction primitives with human sensorimotor abilities. *Interacting with Computers*, 27, 327–343.

Desai, S., Blackler, A., and Popovic, V. (2015). Intuitive use of tangible toys. Paper presented at IASDR 2015, "Interplay," Brisbane, Australia.

Desai, S., Blackler, A., and Popovic, V. (2016). Intuitive interaction in a mixed reality system. Paper presented at DRS 2016, Brighton, UK.

Fischer, S. (2018). Designing intuitive products in an agile world. In A. Blackler (Ed.), *Intuitive Interaction: Research and Application* (pp. 195–212). Boca Raton, FL: CRC Press.

Fischer, S., Itoh, M., and Inagaki, T. (2015a). Prior schemata transfer as an account for assessing the intuitive use of new technology. *Applied Ergonomics, 46* (Part A), 8–20.

Fischer, S., Itoh, M., and Inagaki, T. (2015b). Screening prototype features in terms of intuitive use: Design considerations and proof of concept. *Interacting with Computers,* 27(3), 256–270.

Gudur, R. R. (2012). *Approaches to designing for older adults' intuitive interaction with complex devices.* PhD thesis, Queensland University of Technology, Brisbane, Australia.

Gudur, R. R., Blackler, A., Popovic, V., and Mahar, D. (2009). Redundancy in interface design and its impact on intuitive use of a product in older users. Paper presented at IASDR 2009, International Association of Societies of Design Research Conference, Seoul, South Korea.

Gudur, R. R., Blackler, A. L., Popovic, V., and Mahar, D. (2013). Ageing, technology anxiety and intuitive use of complex interfaces. Proceedings of Human–Computer Interaction: INTERACT 2013. *Lecture Notes in Computer Science, 8119,* 564–581.

Hespanhol, L. (2018). City context, digital content and the design of intuitive urban interfaces. In A. Blackler (Ed.), *Intuitive Interaction: Research and Application* (pp. 173–194). Boca Raton, FL: CRC Press.

Hespanhol, L., and Tomitsch, M. (2015). Strategies for intuitive interaction in public urban spaces. *Interacting with Computers, 27*(3), 311–326.

Hornecker, E. (2012). Beyond affordance: Tangibles' hybrid nature. In Proceedings of the Sixth International Conference on Tangible, Embedded and Embodied Interation (pp. 175–182): Kingston, ON, Canada ACM.

Hurtienne, J. (2009). *Image schemas and design for intuitive use.* PhD thesis, Technischen Universität Berlin, Germany.

Hurtienne, J., and Blessing, L. (2007). Design for intuitive use: Testing image schema theory for user interface design. Paper presented at the 16th International Conference on Engineering Design, Paris, 2007.

Hurtienne, J., and Israel, J. H. (2007). Image schemas and their metaphorical extensions: Intuitive patterns for tangible interaction. Paper presented at TEI '07, the First International Conference on Tangible and Embedded Interaction, New York.

Hurtienne, J., Klöckner, K., Diefenbach, S., Nass, C., and Maier, A. (2015). Designing with image schemas: Resolving the tension between innovation, inclusion and intuitive use. *Interacting with Computers, 27,* 235–255.

Lakoff, G., and Johnson, M. (1980). *Metaphors We Live By.* Chicago, IL: Chicago Press.

Lawry, S., Popovic, V., and Blackler, A. (2010). Identifying familiarity in older and younger adults. Paper presented at the Design and Complexity Design Research Society International Conference 2010, Montreal, Canada.

Lawry, S., Popovic, V., and Blackler, A. (2011). Diversity in product familiarity across younger and older adults. Paper presented at IASDR2011, the 4th World Conference on Design Research, Delft, the Netherlands.

Macaranas, A. (2013). The effects of intuitive interaction mappings on the usability of body-based interfaces. Master's thesis, Faculty of Communication, Art and Technology, School of Interactive Arts and Technology, Simon Fraser University, Burnaby, Canada.

McAran, D. (2018). Development of the technology acceptance intuitive interaction model. In A. Blackler (Ed.), *Intuitive Interaction: Research and Application* (pp. 129–150). Boca Raton, FL: CRC Press.

McEwan, M., Blackler, A., Johnson, D., and Wyeth, P. (2014). Natural mapping and intuitive interaction in videogames. Paper presented at CHI PLAY '14, the First ACM SIGCHI Annual Symposium on Computer–Human Interaction in Play, Toronto, Canada.

McEwan, M. (2017). The influence of naturally mapped control interfaces for video games on the player experience and intuitive interaction. PhD thesis, Queensland University of Technology, Queensland, Australia.

Mihajlov, M., Law, E. L.-C., and Springett, M. (2014). Intuitive learnability of touch gestures for technology-naïve older adults. *Interacting with Computers*, *27*(3), 344–356.MiMff

Mohan, G., Blackler, A. L., and Popovic, V. (2015). Using conceptual tool for intuitive interaction to design intuitive website for SME in India: A case study. Paper presented at the 6th International Association of Societies of Design Research (IASDR) Congress, Brisbane, Australia.

Mohs, C., Hurtienne, J., Israel, J. H., Naumann, A., Kindsmüller, M. C., Meyer, H. A., and Pohlmeyer, A. (2006). IUUI: Intuitive Use of User Interfaces. In T. Bosenick, M. Hassenzahl, M. Müller-Prove and M. Peissner (Eds.), *Usability Professionals 2006* (pp. 130–133). Stuttgart, Germany: Usability Professionals' Association.

O'Brien, M. (2018). Lessons on intuitive usage from everyday technology interactions among younger and older people. In A. Blackler (Ed.), *Intuitive Interaction: Research and Application* (pp. 89–112). Boca Raton, FL: CRC Press.

O'Brien, M., Weger, K., DeFour, M. E., and Reeves, S. M. (2011). Examining the role of age and experience on use of knowledge in the world for everyday technology interactions. Paper presented at the 55th Annual Human Factors and Ergonomics Society Annual Meeting (pp. 171–181). Santa Monica, CA.

O'Brien, M. A. (2010). *Understanding Human–Technology Interactions: The Role of Prior Experience and Age*. Atlanta: Georgia Institute of Technology.

O'Brien, M. A., Rogers, W. A., and Fisk, A. D. (2008). Developing a framework for intuitive human–computer interaction. Paper presented at the 52nd Annual Meeting of the Human Factors and Ergonomics Society, New York.

Palmer, J., Ogunyoka, T., and Hammond, C. (2018). Intuitive interaction in industry user research: Context is everything. In A. Blackler (Ed.), *Intuitive Interaction: Research and Application* (pp. 213–226). Boca Raton, FL: CRC Press.

Reddy, G. R., Blackler, A., and Popovic, V. (2018). Adaptable interface framework for intuitively learnable product interfaces for people with diverse capabilities. In A. Blackler (Ed.), *Intuitive Interaction: Research and Application* (pp. 113–128). Boca Raton, FL: CRC Press.

Reddy, R. G., Blackler, A., Mahar, D., and Popovic, V. (2010). The effects of cognitive ageing on use of complex interfaces. Paper presented at OZCHI 2010, "Design—Interaction—Participation," 22nd Annual Conference on the Australian Computer–Human Interaction Special Interest Group, Brisbane, Australia.

Still, J. D., and Still, M. L. (2018). Cognitively describing intuitive interactions. In A. Blackler (Ed.), *Intuitive Interaction: Research and Application* (pp. 41–62). Boca Raton, FL: CRC Press.

Tretter, S., Diefenbach, S., and Ullrich, D. (2018). Intuitive interaction from an experiential perspective: The intuitivity illusion and other phenomena. In A. Blackler (Ed.), *Intuitive Interaction: Research and Application* (pp. 151–170). Boca Raton, FL: CRC Press.

2 Perspectives on the Nature of Intuitive Interaction

Alethea Blackler, Shital Desai, Mitchell McEwan, Vesna Popovic, and Sarah Diefenbach

CONTENTS

2.1 INTRODUCTION

Intuitive interaction is defined as fast, somewhat non-conscious, and generally accurate interaction with an interface that is informed by past experience or *technology familiarity* (TF) (Blackler, Mahar, & Popovic, 2010). Eighteen years of research into intuitive interaction by various researchers on four different continents using a variety of products, interfaces, and experiment designs has shown that prior experience is the leading contributor to intuitive interaction (Blackler et al., 2010; Fischer, Itoh, & Inagaki, 2014; Hurtienne & Blessing, 2007; Hurtienne & Israel, 2007; Mohs et al., 2006; O'Brien, Rogers, & Fisk, 2008). Other researchers have also conducted experiments on prior experience, especially for older people. They have also found that the similarity of prior experience to the usage situation was the main determinant of fast and error-free interaction (Langdon, Lewis, & Clarkson, 2007; Lewis, Langdon, & Clarkson, 2008). See Chapter 1 (Blackler, 2018) for an overview of the history of intuitive interaction research.

Research in this field has covered applications for physical and digital user interfaces, public spaces, video games, *natural user interfaces* (NUIs), and *tangible and embodied interfaces* (TEIs) for younger and older adults and even children (Blackler & Popovic, 2015). Researchers have also investigated, tested, and provided tools for the most appropriate ways to design more intuitive interfaces (Blackler, Popovic, & Mahar, 2014; Fischer, Itoh, & Inagaki, 2015; Hurtienne, Klöckner, Diefenbach, Nass, & Maier, 2015). However, recent research has suggested both new applications for intuitive interaction and different perspectives for thinking about it. Sections 2.2–2.4 discuss older and newer applications for this research and new perspectives on the experience of intuitive interaction. Sections 2.5 and 2.6 synthesize these ideas into a new framework for intuitive interaction to help us understand how all these contributions interrelate and their implications for researchers and designers.

2.2   EARLY THEORY MAKING IN INTUITIVE INTERACTION RESEARCH

The two groups of early intuitive interaction researchers (in Australia and Germany) developed distinct theories about the types of experiential knowledge accessed during intuitive interaction and how designers could maximize an interface's potential for intuitive use, and there has always been significant overlap between these two approaches (Blackler & Hurtienne, 2007). The German-based Intuitive Use of User Interfaces (IUUI) research group presented a continuum of knowledge in intuitive interaction (Figure 2.1, top) with types of experiential knowledge accessed during intuitive interaction based on their frequency of cognitive encoding and retrieval (Hurtienne & Israel, 2007). The Australian intuitive interaction continuum suggested the means by which intuitive interaction can be supported through design by applying different levels of past experience (Blackler, 2008), shown in Figure 2.1 (bottom) as it relates to IUUI's continuum.

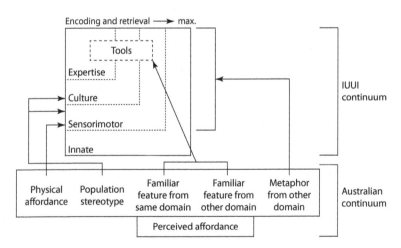

FIGURE 2.1   The IUUI and Australian intuitive interaction continua compared. (Adapted from Blackler and Hurtienne, 2007.)

In IUUI's continuum, the most basic and broadly possessed knowledge identified is innate knowledge, which has genetic origins and manifests in responses such as reflexes. In the Australian continuum, the most accessible design strategy is to use physical affordances, which take advantage of embodied knowledge of the world established early in life. Gibson (1977) stated that individuals perceive what an environment has to offer, in terms of properties, mediums, and compositions, for the possibility of an action. Gibson also emphasized that affordances depend on an individual's ability to perform actions in the environment. A chair, for example, affords sitting for adults, but for infants it is a walker or a support. Norman (1999) referred to Gibson's affordances as *real affordances* or *physical affordances*. Physical affordances are perceptually obvious and do not have to be learned. This fits within IUUI's sensorimotor level, which consists of knowledge from embodied interaction with the physical world, acquired very early in childhood and from then on used continuously through interaction with the world. For example, children learn about gravity and build up concepts for speed. Image schemas also sit at the sensorimotor level as they involve understanding that is built on sensorimotor experience. Image schemas are metaphorical extensions of cognitive concepts that are based on each individual's experience of interaction with the physical world. Because they are based on past experience, and because they are so well known and so universal that they become unconscious, image schemas can be defined as intuitive (Hurtienne, 2009). See Chapter 1 (Blackler, 2018) for a review of image schema research.

Blackler classed the next level of knowledge as *population stereotypes*, which derive largely from experience of cultural conventions (e.g., clockwise to increase, different meanings for hand gestures, or different directions for electrical switches). Population stereotypes relate to IUUI's *culture* and *sensorimotor* levels and include knowledge broadly distributed within specific populations or cultures. When population stereotypes are conformed to, reaction or decision time is shorter, the first movement made is more likely to be correct, the use of a control is faster and more precise, and people learn to use a control more rapidly (Asfour, Omachonu, Diaz, & Abdel-Moty, 1991). The level with the lowest frequency of encoding and retrieval in IUUI's continuum is *expertise*, which is knowledge held only by those adept at a particular specialty. In this case, the term *expertise* is intended to mean both general expertise gained by anyone who has spent time working with a particular tool (or a similar one), as well as the kind of expertise gained by true experts, often defined as requiring 10,000 or more hours of practice (Ericsson, Krampe, & Tesch-Romer, 1993). The separation between the two is implied by the overlay of the *tool* box over the knowledge levels (including *expertise*). At the *sensorimotor* level, there are primitive tools such as sticks for extending one's reach and stones used as weights. At the *culture* level, there are tools commonly used by many people, such as pens and lamps. At the *expertise* stage, there is the knowledge acquired from using tools in one's area of expertise—for example, image-editing tools or computer numerical control (CNC) machines. Even within the same domain of expertise (e.g., graphic design), there may be differing knowledge of tools, depending on the kind of tools previously used (e.g., Corel Paint vs. Adobe Photoshop).

To enable intuitive interaction in the expertise and culture categories, Blackler (2008) suggested using *familiar features* from the *same domain* (e.g., standardized

AV icons such as "Play" and "Stop" used for all AV products) or *differing domains* (e.g., the ubiquitous power symbol used for a new product type). Familiar features tend to be *perceived affordances*, such as an icon button that invites pushing or clicking because a user has learned that that is what it does based on prior experience with similar things (Norman, 2013). Perceived affordance was therefore placed on the Australian continuum as being equivalent to familiar features (Blackler, 2008).

Finally, if the technology or context of use is completely new, then designers can leverage *metaphor* to communicate the intended interaction, to explain a completely new concept or function. Metaphors are grounded in experience (Lakoff & Johnson, 1981) and allow the retrieval of useful analogies from memory and the mapping of the elements of a known situation, the *source*, onto a new situation, the *target* (Holyoak, 1991; Lakoff, 1987). The most obvious successful example is the *desktop metaphor* (Smith, Irby, Kimball, & Verplank, 1982). However, *metaphor* can apply across most of the IUUI continuum from sensorimotor to expertise (e.g., image schemas use metaphorical extensions at the sensorimotor level), and so metaphor is shown in Figure 2.1 linked to sensorimotor, culture, and expertise levels.

The IUUI continuum of knowledge also has an inherent dimensionality. The frequency of encoding and the retrieval of knowledge increases from the top to the bottom of the continuum. Therefore, the closer to the top level of the continuum, the higher the degree of specialization of knowledge and the smaller the potential number of users possessing this knowledge. Similarly, the Australian continuum has a progression in ubiquity from left to right. Physical affordances should be the most ubiquitous interface features a designer can apply, as almost every person should have the embodied experience necessary to understand them. Next come population stereotypes, which are understood by whole populations, then familiar features, which users may or may not recognize depending on their pattern of past experience. Finally, there is metaphor, which may be understood by various people, depending on their experience of the source of the metaphor.

Still and Dark (2010) elicited three types of knowledge from users (*affordance*, *convention*, and *bias*) for the purposes of designing intuitive interfaces for them. Two of these knowledge types corresponded to those on the Australian continuum (affordances = physical affordances; conventions = population stereotypes), and another (which they call bias) may have some equivalence with familiar features. Through their experiment, they have provided empirical evidence for the existence of a continuum of intuitive interaction (Still, Still and Grgic, 2015). Their work is extended and discussed further by Still and Still (2018) in Chapter 3.

This foundation of intuitive interaction theory has formed a solid base that has been applied to design (e.g., Blackler et al., 2014; Hespanhol & Tomitsch, 2015; Hurtienne et al., 2015; Mohan, Blackler, & Popovic, 2015) and built on since. Section 2.3 will discuss different applications of intuitive interaction by various researchers in the areas of natural, tangible, embedded, embodied, and naturally mapped interfaces, and Section 2.4 will discuss how we might start to understand the subjective experience of intuitive interaction.

2.3   EXPANSION OF INTUITIVE INTERACTION APPLICATIONS

TEIs and NUIs have long been claimed to be intuitive (Hurtienne & Israel, 2007; Jacob et al., 2008). This intuitiveness is attributed to tactile or haptic interactions in terms of static system properties such as directness (Dix, 2011), ease of learning and naturalness (Muller-Tomfelde & Fjeld, 2012), and speed, simplicity, and effectiveness (Jacoby et al., 2009). TEIs represent interfaces that accept physical interactions in the form of gestures, touch, and body movements as inputs to the systems. They are comprised of a mix of physical and virtual elements. Depending on the configuration of the physical and virtual elements, TEIs include interfaces that range across a broad spectrum such as *tangible user interfaces* (TUIs) (Ishii, 2008), mixed-reality systems (Milgram & Kishino, 1994), ubiquitous systems (Vallgårda, 2014), and gestural and whole-body systems (Aslan, Primessnig, Murer, Moser, & Tscheligi, 2013). TEIs with gestural interaction and body movements were made popular in futuristic movies such as *Star Trek* and *Minority Report*. Their use has been successfully implemented in gaming environments such as Microsoft's Xbox Kinect and Nintendo's Wii Remote. Other TEIs such as Siftables (Merrill, Kalanithi, & Fitzgerald, 2011) and reacTable (Jordà, Geiger, Alonso, & Kaltenbrunner, 2007) embed physical artifacts with digital information so that the information can be directly manipulated and accessed (Ishii, 2008). Similarly, physical and virtual elements coexist in a mixed-reality system (Milgram & Colquhoun, 1999) such as Osmo (Tangible Play, 2014). Although TUIs and mixed-reality systems are referenced separately in the literature, there is no clear way to differentiate between them. The commonality between them is that they both integrate physical and virtual elements and they both have two main categories of implementation: (1) distinct and separate physical and virtual spaces and (2) overlapping physical and virtual spaces.

If placed on a physical–virtual continuum with physical products (e.g., blocks to play a game of Jenga; Hasbro & Scott, 2001) on the extreme left and virtual interfaces (e.g., software) on the extreme right, the rest of the continuum is occupied by TEIs. TEIs with more physical interactions are placed toward the left, while TEIs that allow more interactions with virtual interfaces are placed toward the right. TEIs involve more everyday movements, interactions, and gestures than many more traditional interfaces, which theoretically should place them at the lower end of the intuitive interaction continua (Figure 2.1). For example, they use physical affordances such as touching and grasping, innate responses such as turning toward a stimulus, population stereotypes such as shaking the head, and sensorimotor actions such as moving up and down. Their enhanced intuitive potential is possibly due to these interfaces accessing sensorimotor-based knowledge, which primarily compose image schemas (Hurtienne & Israel, 2007), and their use of more literal "reality-based" interaction mappings (Jacob et al., 2008).

However, the assumed increased intuitiveness of TEIs had not been empirically shown until recently. Desai et al. (2015) compared the use of a physical toy (physical Jenga) with an equivalent virtual interface (Jenga app) by children aged 5–11 and found that the physical toy was more intuitive and gameplay was more successful than with the equivalent virtual app. Intuitive interactions with the physical toy were facilitated by its high level of reliance on physical affordances, as opposed to a reliance on perceived affordances in the virtual game. In another experiment with a TEI for children, Desai, Blackler, and Popovic (2016) found that there were more intuitive interactions than non-intuitive interactions coded in the audiovisual data. In addition, linear regression analysis established that the variation in intuitive interaction in the TEI could be statistically explained primarily by physical affordances offered by the TEI and to a lesser extent by the perceived affordances in the system. Therefore, physical affordances are primary contributors to intuitive interaction in TEIs that utilize tactile interactions with physical elements. However, Desai et al. (2016) also raised the issue that the intuitive interaction in TEIs heavily depends on how the physical and virtual elements are configured in the system, and so intuitiveness cannot be assumed for any TEI but depends on good design; intuitive interaction in TEIs is not a given just because an interface has physical elements. For intuitive interaction to occur, the right pathways need to be provided, such as physical affordances.

A complimentary approach to the classification of NUIs and TEIs is to assess the extent that they employ *natural mapping*. Natural mapping utilizes physical analogies and cultural standards to communicate a clear relationship between a control interface and its actions in the system (Norman, 2013). The level of natural mapping in the interface is thus determined by the degree of correspondence between physical control inputs and their system responses. A spatial analogy in the interface could utilize directional correspondence or might employ groups or patterns that reflect the relationship between controls and feedback. For example, using cursor keys to control movement on a computer or aligning physical function keys with their corresponding interface elements on a screen (e.g., pre-touchscreen phones, cameras, and similar devices). Natural mapping is also represented in Blackler's (2008) original Australian continuum, where it is referred to as *compatible mapping*, drawing on population stereotypes (and to some extent physical affordances) to provide a consistent and familiar interface layout to users. In this way, naturally mapped interfaces can also leverage sensorimotor and cultural knowledge established through interaction with other systems or real-life activities—for example, swinging a tangible controller as though it were a golf club while playing a golf video game.

McEwan and colleagues categorized a range of control interfaces for racing and tennis video games according to their level of natural mapping, including many that could be classed as TEIs (McEwan, 2017; McEwan, Blackler, Johnson, & Wyeth, 2014). Categorization was undertaken with an existing typology for naturally mapped control interfaces (NMCIs) for video games (Skalski, Tamborini, Shelton, Buncher, & Lindmark, 2011). McEwan et al. used a version of the *technology familiarity* (TF) questionnaire previously used by Blackler et al. (2010) and revised to measure relevant previous video game experiences to create a *game technology familiarity* (GTF) score, as well as employing empirical objective and subjective measures of intuitive use. McEwan (2017) was able to show that GTF broadly predicted objective measures of intuitive use, explaining variations revealed by other demographic factors such as age and gender. For example, the oldest group consistently showed significantly lower intuitive use outcomes than the youngest group across NMCIs, and these differences were also shown in their GTF score for five of the six NMCIs tested. However, the control interfaces with increased natural mapping also showed compensatory effects, especially for participants with reduced familiarity, that generally allowed for equivalent or increased intuitive interaction. The more naturally mapped interfaces were also consistently subjectively perceived to offer more intuitive controls. McEwan (2017) identified a range of factors that could contribute to naturally mapped controls supporting intuitive use, such as their level of physical realism, their bandwidth (or fidelity), and their naturalness (or the way control actions are naturally mapped between physical and virtual representations). Both increased natural mapping, and greater domain familiarity may therefore lead to more intuitive interactions with video games.

Antle, Corness, and Droumeva (2009a) used embodied conceptual (image schematic) metaphors in the design of a playful sound-making full-body interaction interface and concluded that discoverability through feedback and tight mapping are prerequisites for embodied intuitive interaction. They compared an embodied interface using two different control mappings: one using conceptual metaphors (image schemas) and the other using more random mappings. Their main finding was that simply using image schemas is not enough to make the interface intuitive; the control actions (and resulting system response) should also be easily discoverable by the users. This implies that, in order to allow knowledge transfer and increase the intuitive potential of embodied interfaces that rely on more than just a digital input or a feature, the actual control actions may have to be obviously highlighted for users. With a full-body interaction TEI, it is not just features in the interface that need considering but the user's actions and movements in the whole context; that is, it is not just pressing a button but *how* to perform the action *in context*.

Macaranas, Antle, and Riecke (2015) described an experiment that tested three different full-body gestural interfaces to establish which types of mappings were more intuitive, one based on image schemas and two on different previously encountered features from other types of interfaces. Macaranas et al. (2015) asked their participants about how well they understood both the operation of the system they had used during their experiment and the content presented through that system. They found that intuitiveness as measured by performance was not all that users wanted from a system. For example, if they did not discover the interaction model behind the controls, they felt dissatisfied. On the other hand, where the underlying concept was not so easily discovered, users tended to engage more with the content presented through the system. Macaranas et al. (2015) therefore suggested that an unconscious (rather than conscious or explicit) understanding of the system enabled participants to focus their conscious attention on completing the tasks, not on learning to use or using the interface. Their paper highlighted the complex mix of requirements that entertainment-based systems have; ease of use and intuitiveness may not always be as important as immersion, engagement, or challenge in these types of environments.

However, in an experiment with an augmented and tangible storybook system designed for children, Hornecker (2012) found that the physical properties of TEIs created unconscious expectations for the users that resulted in a mismatch between the system's conceptual model for interaction and the mental model assumed by users. Rather than purely focusing on facilitating unconscious knowledge transfer and "natural" interaction, Hornecker argued that these systems should be *seamful*, offering opportunities for conscious reflection and learning that adjusts expectations to match the possible inputs with the resulting system response.

Antle, Corness, and Droumeva (2009b) and Macaranas et al.'s (2015) findings that users may misread system feedback or affordances to incorrectly assume that certain control actions were accepted as correct by the system (and continue to use them in error), as well as Hornecker's (2012) findings, suggest that NUIs are by no means a silver bullet. Norman (2010) also offered criticism, suggesting that NUIs are neither natural nor easy to learn or remember. Some gestures could be confusing as they differ from one culture to another—for example, Indian head-shaking and hand-waving gestures, which often differ in meaning from Western ones. Derivation of meaning from gestures and body movements relies heavily on an individual's past experience and knowledge, so careful design taking into account users' culture is essential. In addition, Chattopadhyay and Bolchini (2015) argued that embodied interfaces that utilize 3D gestures are drawing on expertise rather than sensorimotor knowledge, limiting their intuitive potential, and proposed a 2D gesture system based on image schemas to resolve this problem. They also found that biomechanical factors influenced results, with ergonomics and gesture position and length playing a role in the accuracy and ease of execution of gestural strokes.

As a whole, the literature shows that NMCIs, NUIs, and TEIs have the potential to make a system more intuitive but suggests that we need to design these systems carefully. They can indeed offer great opportunities for intuitive interaction, yet there are unique design challenges that must be overcome—for example, providing pathways such as physical affordances, applying the correct mapping, making appropriate interactions discoverable, and considering ergonomic, aesthetic, and technological factors. In short, adding a "natural" user interface to a system does not automatically increase intuitive potential, but these findings suggest that through considered design, a natural, embedded, embodied, tangible mixed-reality or naturally mapped user interface can improve intuitiveness.

2.4 SUBJECTIVE PERSPECTIVES ON INTUITIVE INTERACTION

This section goes beyond looking at where and how intuitive interaction has been applied and explores some new perspectives on the nature of intuitive interaction itself that have placed more focus on the subjective experiences of users. In earlier work, objectively measured intuitive uses were used in research more than subjectively rated ones, but more recently, researchers have begun to look more seriously at user self-reporting of perceived intuitiveness (e.g., Diefenbach & Ullrich, 2015; McAran, 2018; McEwan et al., 2014; Tretter, Diefenbach, & Ullrich, 2018).

The subjective experience of an intuitive interaction was identified in the early Australian work as potentially related to the subjective feelings that accompany an intuitive insight, or the "Aha!" moment, yet was not further explored (Blackler, 2008). According to research in decision making, the "Aha!" experience (also called *insight restructuring*) is an instinctive affective response involving suddenness (Bowden, 1997). Bowden's (1997) experiments tested the hypothesis that the "Aha!" experience associated with insight solutions is related to unreportable processing; it was hypothesized that unreportable memory activation and retrieval processes can make the solution available but leave the person unaware of its source. They showed that unreportable hints during problem solving led to insight-like solutions, which provided evidence that unreportable processing can produce the "Aha!" experience of insight solutions. It could be concluded from this that intuition may be the unreportable, unconscious process that leads to insight. The environment and the stored knowledge combined provide the clues needed for a successful solution (Bowden, 1997), just as intuition combines information in the situation and in memory to come up with solutions (Bastick, 2003). The "Aha!" experience can then be seen as the affective product of the unconscious processing.

Still et al. (2015) called for more research into these affective aspects of intuitive interaction, highlighting that affect is repeatedly identified as a core component of intuitive experience and has been clearly linked to familiarity. Diefenbach & Ullrich (2015) found further evidence to support the link by taking a new approach to defining intuitive interaction focused on the users' subjective interpretation of the phenomenon. Through their work, also extended in Chapter 8 (Tretter et al., 2018), four components of the subjective experience of intuitive use were identified: gut feeling (guided by unconscious thought), verbalizability (hard to identify the source of insight), effortlessness (quick and easy interaction), and magical experience (accompanied by feelings of positive affect), as summarized in their INTUI framework. They developed a standardized INTUI questionnaire to measure these four components of intuitive interaction. They further theorized that the *domain transfer distance*, "the distance between the application domain and the origin of prior knowledge that enables intuitive interaction" (Diefenbach & Ullrich, 2015, p. 218), interacts with these components to influence perceptions of intuitive use. If the domain transfer distance is low (as with image schemas and sensorimotor knowledge), then effortlessness will be maximized in the experience and the other components will be minimized. If the domain transfer distance is high (as with features transferred from distant domains or with the use of metaphor), then gut feeling and magical experience will be maximized in the subjective experience (up to a theorized limit where the potential for knowledge transfer and intuitive use fade), while the other components are minimized.

Through sampling user ratings of intuitiveness and the four INTUI components using hypothetical usage scenarios, Diefenbach and Ullrich (2015) were able to offer some validation of the theorized relationships between their constructs, opening up a new avenue of exploration for intuitive interaction researchers. They found that there was a high level of agreement about the four components of their model and also that participants judged scenarios with a higher transfer distance as more appropriate representations of intuitive interaction. In other words, participants saw magical experience and gut feeling, which are the subjective experiences of high transfer interaction, as more typical of intuitive interaction than effortlessness and verbalizability, which are the kinds of objective experiences generally coded as intuitive interaction in previous research. Of course, these can also potentially overlap, as an objectively measured intuitive interaction could also be subjectively magical for the user.

It is interesting that Macaranas et al.'s (2015) and Antle et al.'s (2009a) findings on discoverability and transparency have some similarities with Diefenbach & Ullrich's (2015) investigation into the subjective experience of intuitive interaction and with Hornecker's (2012) call for more seamful interfaces. The magical or mysterious experiences delivered by more implicit knowledge, where the source of the knowledge is not recalled, could be a promising avenue to deliver more engaging yet intuitive interfaces, but Macaranas et al.'s (2015) findings suggest that, for some applications, the experiences delivered by the options in the center of the continua, where users may well have consciously "discovered" their origin by the end of the interaction, are a safer option for providing a usable interface.

This suggests that where on the continua the prior knowledge sits affects the subjective experience; for example, physical affordance (sensorimotor) and even population stereotypes (culture) could be so engrained that they are unconscious, feel automatic, and go unnoticed by the user, as suggested by the work of Still et al. (2015) on what they refer to as *biases* and *conventions*, both of which appeared to be largely unconsciously applied by their participants (Still & Still, 2018). On the other hand, metaphor, if done right, offers a potential route for increasing domain transfer distance and designing more subjectively magical experiences. A feature with higher transfer distance could appear more mysterious because users may not consciously remember where their knowledge about it came from. Hence, because it is implicitly known and somewhat unexpected in the context, it appears more magical. In between, familiar features may make for a more measureable but possibly more pedestrian experience.

2.5 ENHANCED FRAMEWORK FOR INTUITIVE INTERACTION (EFII)

These newer ideas and approaches have exciting potential to expand the field of intuitive interaction and to inform the designers of a variety of systems about how to make interfaces both engaging and intuitive. However, we need to comprehend how they relate to each other if we are to have a coherent understanding of how to apply intuitive interaction going forward. Building on work done in the past that compared and contrasted the two separate continua of intuitive interaction (Blackler & Hurtienne, 2007; Figure 2.1), an enhanced framework to explain how these newer ideas relate to older ones is shown in Figure 2.2. This enhanced framework for intuitive interaction (EFII) is intended to offer an aid to understand how newer concepts relate to ideas in the existing continua. Here, the previously developed Australian continuum (Blackler, 2008; Figure 2.1, bottom) is reconceptualized as *Pathways to intuitive use* and aspects of the IUUI continuum (Figure 2.1, top) are included in the *Knowledge* bar at the bottom. Findings from the more recent work on *Interface types*, *Characteristics of features*, and sources of *Knowledge* in intuitive interaction are mapped onto this framework to suggest the means by which they can be understood in relation to each other and applied to the *Pathways to intuitive use* to resolve design problems and increase intuitiveness. The framework highlights parallels and connections between the different dimensions of intuitive interaction shown on the left side: *Pathways to intuitive use*, *Interface types*, *Characteristics of features*, and the source of the *Knowledge* that enables intuitive use. Note, however, that the connections between the different dimensions do not represent fixed relationships, nor do concepts on opposing poles necessarily represent contradictions or exact opposites. For example, *Magical* is not necessarily opposed to *Image schemas*, and although it may be non-compatible with *Transparent* due to the different sources of prior experience, it is not an actual opposite.

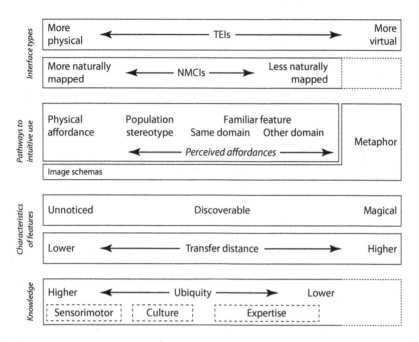

FIGURE 2.2   Enhanced Framework for Intuitive Interaction (EFII).

In this enhanced framework, *Metaphor* has been detached from the other parts of the original Australian continuum. This is because it has become clear that is it not always a simple continuation from the other concepts and in fact could be applied in other ways than originally assumed. The extension of the *Metaphor* block beneath the other *Pathways to intuitive use* is intended to demonstrate that metaphor can in fact be applied through any of the other pathways (e.g., physical affordances, familiar features, etc.). For example, the metaphor of pinch to zoom on touch devices originated from the idea of stretching a surface such as a balloon, but it is now applied as a familiar feature since it is so widespread. Tilt and gesture controls, based on physical affordances and sensorimotor knowledge, are also now commonly used in smart devices and games, and users may transfer knowledge straight from other similar devices rather than from the original sources of the metaphors. Hurtienne et al. (2015) also showed that image schemas, which sit in this *Metaphor* spur because they use metaphorical extensions of basic cognitive concepts, can be ubiquitous (or known to most people in a society) and hence intuitive and inclusive, belonging on the left side of the framework with higher ubiquity and sensorimotor knowledge rather than on the right with more traditional metaphors.

Thus, *interface types* that are more physical mostly rely on physical affordances that enable grasping, holding, and sliding physical objects. They can, however, also leverage population stereotypes (or cultural conventions) and metaphors (e.g., turning a wheel, using a racket to play tennis, balancing blocks one above the other to create a stack). Likewise, more naturally mapped control interfaces are likely to leverage physical affordances and population stereotypes to enable intuitive interaction through embodied knowledge. On the far right of the natural mapping bar, the empty dotted box refers to either arbitrary (i.e., random and not natural) mapping or mappings that leverage metaphor to potentially access other areas of the pathways to enable intuitive use.

As the TEIs become more virtual or the NMCIs become less naturally mapped (toward the right of *Pathways to intuitive use*), the interfaces rely on the other pathways, depending on the system and its design. In these cases, transfer distance increases and ubiquity decreases, making it more difficult for designers to utilize natural mapping to increase the intuitive potential of their interfaces. Less naturally mapped interfaces are therefore more likely to rely on familiar features and metaphor as pathways to intuitive use. This is generally the case with more complex systems that cannot be simply mapped at the sensorimotor level (e.g., software or menu-driven role-playing games). In these cases, the level of previous domain experience (or TF) is more relevant in determining the intuitive potential of the system, though again, metaphor may be used as a bridge to more ubiquitous Knowledge. For example, Hurtienne (2009) used image schemas to increase the intuitiveness of accounting software features. In line with this, mixed TEI systems (including physical and virtual elements) could access all parts of the *Pathways to intuitive use*, depending on the proportion of physical and virtual elements in the system, although such a design in one single system is perhaps unlikely and may be strained in communicating a clear conceptual model.

Magical experiences appear to relate to increased transfer distance and are therefore most likely to be induced by metaphors. A metaphor may appear magical because users may be able to apply the metaphor but be unable to recall the source of their *Knowledge*. At the other end, physical affordances appear to facilitate transparent interactions that could be delivered with or without the application of image schemas. Very simple and engrained knowledge such as physical affordances could pass unnoticed as it is so well used and so expected. Vera and Simon (1993), Baber and Baumann (2002), and Norman (1993) all agree that tools that stay in the background allow the user to directly engage with the task. For example, "The most profound technologies are those that disappear. They weave themselves into the fabric of everyday life until they are indistinguishable from it" (Weiser, 1991, in Baber and Bauman, 2002, p. 285). Until a system stops working, many people would not be consciously aware of the properties of features at the left end of *Pathways to intuitive use*; they are transparent and go unnoticed, remaining unconscious. Discoverable experiences seem likely in the center part of *Pathways to intuitive use*, where users are most likely to recognize the previous knowledge they are applying. However, discoverable does not always mean discovered; whether particular users consciously recognize the interface elements will depend on their relevant past experience and possibly their level of engagement with the particular task.

Finally, the ubiquity of previous experience and potential for more people to be able to intuitively use a feature are highest at the lower end of *Pathways to intuitive use* and decrease from left to right. *Metaphor* is again a potential exception here as a very universal metaphor could be applied in some cases (e.g., 'up to increase', 'progress as travel along a path', which are both image schemas). It should be noted that there will be exceptions to these examples, and this framework does not present hard-and-fast rules. For example, Desai et al (2015) did find some use of perceived affordances with a physical toy, but most overall uses and intuitive uses were facilitated by physical affordances. Similarly, physical affordances and metaphors could both be discoverable, but they may be less discoverable by most people than perceived affordances due to their extreme familiarity and transparency in the case of physical affordances and their more obscure knowledge base and higher transfer distance on the part of metaphor.

The EFII has some overlaps with the hierarchy presented by Still and Still (Chapter 3, 2018), but instead of aligning the concepts of cognitive science and intuitive interaction, it is intended to explain all the various concepts developed in intuitive interaction theory and to allow designers to apply these ideas to their interfaces through the *Pathways to intuitive use*.

2.6 DISCUSSION

This section discusses some examples and implications of EFII, taking the concepts (bars) from the top to the bottom of Figure 2.2. The *Pathways to intuitive use* and aspects of the IUUI continuum are not discussed separately but instead referenced throughout.

2.6.1 Tangible, Embedded, and Embodied Interfaces

Physical toys such as Jenga and LEGO (2014) offer physical affordances to grasp, hold, push, and pull because of their natural spatial and material properties. The physical properties of the blocks (e.g., the spatial alignment of the blocks in LEGO) are perceived and mapped onto decisions and actions (e.g., to stack them or attach them together). Tactile interactions with these physical products require prior experience playing with similar toys or knowledge from everyday life. Intuitive interaction with physical products thus uses physical affordance and some perceived affordances. In other words, physical products use the lower end of the *Pathways to intuitive use* continuum, as there is a predominant use of physical affordances. This means that there is minimal complexity required in the intuitive use of such physical products and maximum ubiquity in terms of the percentage of people who will be able to access them. Physical toys such as Jenga and LEGO are associated with low domain transfer distance, as the origin of prior knowledge and the application of that knowledge relates to the same physical domain with spatial and material characteristics.

Virtual interfaces, such as an app for the game of Jenga, have limited real physical affordances, thus relying on users' past experience. People use cultural conventions associated with touchscreens, such as swiping the screen, and also conventions pertaining to the virtual representation of physical objects in the app (e.g., blocks). The intuitive use of virtual interfaces is thus associated with population stereotypes (e.g., red for danger included in the design of app features) and perceived affordances (e.g., swiping). The intuitive use of virtual interfaces is also associated with higher domain transfer distance as the prior knowledge acquired from the physical domain (e.g., pushing and pulling the block from the stack) is applied to a virtual domain (e.g., swiping at the touchscreen). The transfer of prior knowledge from the origin to the application domain is often indirect, so it is less ubiquitous as it depends on specific past experience.

2.6.2 NATURALLY MAPPED CONTROL INTERFACES

For NMCIs, the level of natural mapping determines the *Pathways to intuitive use, Characteristics of features,* and the type of *Knowledge* leveraged. For example, NMCIs in a racing video game might range from a realistic arcade-style racing wheel controller to a traditional dual-analogue stick controller (as explored in McEwan, 2017). The racing wheel is more naturally mapped since it has higher realism, bandwidth, and naturalness in terms of the correspondence between the control interface and the real-life activity (racing) that is simulated by the game. The racing wheel leverages physical affordances and population stereotypes (sensorimotor and cultural *Knowledge*) since it is shaped and manipulated like the equivalent real-life interface (a steering wheel). This may cause features to be transparent and not consciously recognized, since the mapping is literal and transfer distance is low. In contrast, the traditional Xbox 360 controller uses familiar features to help players control the car (using directional correspondence through the analogue sticks), relying more heavily on less ubiquitous *Knowledge* (or higher GTF). This means that transfer distance is higher and the source of intuitive interaction may be more discoverable due to the mental effort required. To some extent, NMCIs for video games always leverage metaphor; the correspondence of natural mapping to the simulated activity is a type of metaphor, even if the natural mapping is high. As such, magical experiences may be possible with any control interfaces that leverage natural mapping, yet greater use of metaphor may also increase transfer distance and require a higher level of TF for intuitive use. In all, higher natural mapping in the control interface provides another tool that is complementary to prior experience and can compensate for a lack of relevant familiarity.

2.6.3 DISCOVERABILITY, TRANSPARENCY, AND MAGIC

Finding the right balance between discoverability, intuitive use, and subjective feelings of competence and satisfaction is a challenge that needs meeting. It is possible that reducing the transfer distance could increase intuitive interaction but in some cases also reduce satisfaction, as suggested by Macaranas et al. (2015). Knowledge of some features could be so engrained they are not consciously noticed (they are transparent, like many physical affordances at the bottom end of the *Pathways to intuitive use*), or the metaphor is so seamless its source is not consciously realized, and so the resulting interaction appears magical (at the right side of the *Pathways to intuitive use*). However, not every feature will have a perfect metaphor as not every function has a neatly applicable source and/or target for metaphor, so some metaphors may be less smooth and more likely to be brought to a users' consciousness.

In addition, there may be many options to allow "magical experiences," be it with tangibles or digital objects. For example, if a user lifts up a tangible and it multiplies the corresponding virtual representation, whereas placing it back down reduces it, it uses an image schema, but it applies it in a non-literal way. If gestures or movement are translated into system changes in a consistent yet non-literal way, then the interaction may be perceived as magical—for example, Antle's (2009b) interface where participants did not understand why the changes were happening but could reproduce them through consistent interaction, or interacting with large public interactive art displays that cause changes in the system (Hespanhol, 2018) can appear magical.

Ideally, researchers could develop ways in which designers can provide both magical (subjectively intuitive) and unconscious (objectively intuitive) types of experiences for ultimate ease of use and engagement. While earlier approaches mainly focused on how quantifiable prior knowledge contributes to intuitive interaction, performance parameters, and related design principles, Diefenbach and Ullrich's (2015) approach puts more emphasis on the subjective experience of intuitive interaction and its different facets. A better understanding of the differences between these two categories of intuitive uses (subjective and objective) would allow researchers to develop ways in which designers can create experiences that are subjectively engaging ("magical," as assessed by participants) as well as objectively simple and easy to understand ("intuitive," as coded by researchers) by using the right combination of features in an interface.

2.6.4 TRANSFER DISTANCE

Transfer distance is an important concept because of its potential to allow designers to fairly simply understand how best to apply intuitive interaction to various interface features and also because manipulating transfer distance in more complex ways may allow users to have more magical experiences. The resemblance of the purchase procedure between online and offline shops (browsing items, putting them to a shopping basket, seeing the sum of purchase, making a choice how to pay) could be considered a case of a metaphor with low transfer distance. A procedure known from offline shopping is transferred to online shopping. In both cases, the context is shopping, and parallels between online and offline are highlighted and made obvious to the user. In consequence, the online shopping procedure should appear intuitive but probably not especially magical. An example of higher transfer distance and more magical experience of intuitive use could be a TV remote that controls a cross-shaped cursor, similar to those popular in the context of gaming consoles such as the Nintendo Wii but not yet all that common on TV controls. Both (TV and gaming console) are in the domain of entertainment electronics, but the use contexts are not as close as in the shopping example. Finally, think of an elevator where closing the door can be sped up by repeatedly pressing a button. Performing an action repeatedly to speed up a reaction is a natural pattern of human interaction in other domains (e.g., we say "Hurry up, hurry up" or clap to make sure others understand the urgency), so the transfer distance is relatively high, and if an elevator responded in this way, people might find it somewhat magical.

In the widest sense, the concept of transfer distance could also be considered where elements of positive natural practices with real objects are transferred to other contexts and interaction with digital objects. For example, Diefenbach et al. (2017) transferred elements of *secretive interaction*, identified in peoples' natural secrecy practices in their private lives, to interaction with a digital picture frame in an office setting. The picture frame, besides the public, visible picture, holds a secret picture that may be consumed in a private moment and act as a source of positive feelings of autonomy. Touching the public picture that is visible to all reveals the hidden picture underneath. Concealing this is fast and instantaneous; quickly removing the finger from the screen is a typical gesture when in danger of being caught in the act (Diefenbach et al., 2017). Even though the physical interaction (touching a screen) is nothing new, the interaction may appear magical, since it transfers an interaction known with physical objects into a new context and connects it with a new object.

2.6.5 UBIQUITY

As the examples given in this and the previous section have shown, ubiquity (or near ubiquity) can be achieved by applying appropriate interface features that relate to the lower (left-hand) end of the *Pathways to intuitive use*, to ensure that features are known to everyone or almost everyone in a target population. Features based on physical affordances, image schemas, and population stereotypes should be more intuitive to use for more people. As Fischer (2018) explains in Chapter 10, such familiar things allow designers quite a lot of flexibility in terms of look and feel because they are so familiar, and the stereotypes remain robust even if they are tweaked or adapted to fit a certain aesthetic. These are the things that are so familiar they become transparent until they break down or are designed away from their origins and break the mold (i.e., are redesigned so much that they are no longer recognizable, no longer fit the stereotype, or possess the affordance). More ubiquitous interface features are generally required for things that are accessed by large and diverse numbers of people who are not able to be trained in their use and may use them infrequently or once only (e.g., ticket-vending machines). Making these sorts of things intuitive for everyone means systems will run more smoothly as delays can be reduced.

Ubiquity is important in the application of intuitive interaction as there are many interfaces that are intended to be used by almost anyone. Vending machines have been mentioned, but there are many more examples: point-of-sale systems, ATMs, websites of various types (e.g., banking, patient records, and government information), operating systems for phones, tablets, and computers, and even the installations described by Hespanhol (2018) in Chapter 9. Using many of these interfaces has become less of a choice and more of a necessity in recent years, as everyone is expected to be able to access and manipulate their money and information online, and people need to be able to communicate using tools such as email, social media, and text if they are to stay connected with the rest of society. More specialized tools and equipment can utilize some features from the lower end of the *Pathways to intuitive use* continuum (e.g., power symbol, color codes, image schemas) but at some point will require relevant familiar features and even metaphor to be applied. McAran (2018) has made progress in investigating how these types of specialized tools might be perceived as intuitive and accepted into people's working lives.

2.6.6 SUMMARY

This discussion has woven the threads between the dimensions of intuitive interaction in EFII, using examples to show how each of them relates to the others and to explore how designers could utilize EFII to make their interactions intuitive, ubiquitous, magical, and/or transparent according to the needs of their users. For example, criticisms leveled at TEIs relating to confusion of culturally specific gestures and lack of discoverability of interface workings, leading to lack of satisfaction, can be mitigated by the appropriate use of this framework. The design of gestures taking into account users' culture is facilitated, as is decision making about the appropriate level of transparency or discoverability. Therefore, more intuitive interfaces with the correct cultural references and the desired level of discoverability, transparency, or magic are made possible.

2.7 CONCLUSION

This chapter has provided an overview of newer directions and concepts in intuitive interaction research and brought together the disparate ideas into an enhanced framework in order to foster better understanding of the various concepts. The EFII combines all of the newer ideas within the context of the established work to help clarify understanding about what intuitive interaction is in all its incarnations and how it can be facilitated. The framework is based on four different dimensions; *Interface types, Pathways to intuitive use, Characteristics of features, and Knowledge.* It offers insights into how these perspectives relate to each other. This insight can aid understanding for researchers as well as designers who want to make interfaces more intuitive. Furthermore, the framework will hopefully inspire further research on the relationships between the different dimensions of intuitive interaction and thereby add to a stronger integration of knowledge and different strands of research in this field.

This framework will allow designers to apply the various ideas around intuitive interaction with more confidence and better clarity, and researchers to build on the extant work in the intuitive interaction field, to develop it further, and offer more comprehensive tools and recommendations to designers. It also facilitates a shift in focus from intuitive processes and interfaces to an experience-oriented view on intuitive interaction, while still allowing the original foci to remain relevant. This in turn will allow future studies on ways to incorporate those new approaches to design and assist researchers in understanding how to make interfaces both intuitive and engaging for the widest range of users.

REFERENCES

Antle, A. N., Corness, G., & Droumeva, M. (2009b). Human–computer intuition? Exploring the cognitive basis for intuition in embodied interaction. *International Journal of Arts and Technology*, 2(3), 235–254.

Antle, A. N., Corness, G., & Droumeva, M. (2009a). What the body knows: Exploring the benefits of embodied metaphors in hybrid physical digital environments. *Interacting with Computers*, 21(1–2), 66–75.

Asfour, S. S., Omachonu, V. K., Diaz, E. L., & Abdel-Moty, E. (1991). Displays and controls. In A. Mital & W. Karwowski (Eds), *Workspace Equipment and Tool Design* (pp. 257–276). New York: Elsevier.

Aslan, I., Primessnig, F., Murer, M., Moser, C., & Tscheligi, M. (2013). Inspirations from honey bees: Exploring movement measures for dynamic whole body gestures. Paper presented at the 2013 ACM International Conference on Interactive Tabletops and Surfaces, St Andrews, UK.

Baber, C., & Baumann, K. (2002). Embedded human computer interaction. *Applied Ergonomics*, 33(2002), 273–287.

Bastick, T. (2003). *Intuition: Evaluating the Construct and Its Impact on Creative Thinking*. Kingston, Jamaica: Stoneman and Lang.

Blackler, A. (2008). *Intuitive Interaction with Complex Artefacts: Empirically-Based Research*. Saarbrücken, Germany: VDM.

Blackler, A. (2018). Intuitive interaction: An overview. In A. Blackler (Ed.), *Intuitive Interaction: Research and Application* (pp. 3–18). Boca Raton, FL: CRC Press.

Blackler, A., & Hurtienne, J. (2007). Towards a unified view of intuitive interaction: Definitions, models and tools across the world. *MMI-Interaktiv*, 13(2007), 37–55.

Blackler, A., Mahar, D., & Popovic, V. (2010). Older adults, interface experience and cognitive decline. Paper presented at OZCHI 2010, "Design—Interaction—Participation," 22nd Annual Conference of the Australian Computer–Human Interaction Special Interest Group, Brisbane, Australia.

Blackler, A., & Popovic, V. (2015). Towards intuitive interaction theory. *Interacting with Computers*, 27(3), 203–209.

Blackler, A., Popovic, V., & Mahar, D. (2014). Applying and testing design for intuitive interaction. *International Journal of Design Sciences and Technology*, 20(1), 7–26.

Bowden, E. M. (1997). The effect of reportable and unreportable hints on anagram solution and the Aha! experience. *Consciousness and Cognition*, 6, 545–573.

Chattopadhyay, D., & Bolchini, D. (2015). Motor-intuitive interactions based on image schemas: Aligning touchless interaction primitives with human sensorimotor abilities. *Interacting with Computers*, 27, 327–343.

Desai, S., Blackler, A., & Popovic, V. (2015). Intuitive use of tangible toys. Paper presented at IASDR 2015, "Interplay," Brisbane, Australia.

Desai, S., Blackler, A., & Popovic, V. (2016). Intuitive interaction in a mixed reality system. Paper presented at DRS2016, Brighton, UK.

Diefenbach, S., Hassenzahl, M., Eckoldt, K., Hartung, L., Lenz, E., & Laschke, M. (2017). Designing for well-being: A case study of keeping small secrets. *The Journal of Positive Psychology*, 12(2), 151–158. doi.org/10.1080/17439760.2016.1163405

Diefenbach, S., & Ullrich, D. (2015). An experience perspective on intuitive interaction: Central components and the special effect of domain transfer distance. *Interacting with Computers*, 27(3), 210–234.

Dix, A. (2011). Physical creatures in a digital world. Paper presented at the 29th Annual European Conference on Cognitive Ergonomics, Rostock, Germany.

Ericsson, K. A., Krampe, R. T., & Tesch-Romer, C. (1993). The role of deliberate practice in the acquisition of expert performance. *Psychological Review*, 100(3), 363–406.

Fischer, S. (2018). Designing intuitive products in an agile world. In A. Blackler (Ed.), *Intuitive Interaction: Research and Application* (pp. 195–212). Boca Raton, FL: CRC Press.

Fischer, S., Itoh, M., & Inagaki, T. (2014). Prior schemata transfer as an account for assessing the intuitive use of new technology. *Applied Ergonomics, 46*(2015), 8–20.

Fischer, S., Itoh, M., & Inagaki, T. (2015). Screening prototype features in terms of intuitive use: Design considerations and proof of concept. *Interacting with Computers, 27*(3), 256–270. https://doi.org/10.1093/iwc/iwv002.

Gibson, J. J. (1977). The theory of affordances. In R. Shaw & J. Bransford (Eds.), *Perceiving, Acting, and Knowing* (pp. 67–82). Hillsdale, NJ: Lawrence Erlbaum.

Hasbro & Scott, L. (2001). Jenga. Parker Brothers.

Hespanhol, L. (2018). City context, digital content and the design of intuitive urban interfaces. In A. Blackler (Ed.), *Intuitive Interaction: Research and Application* (pp. 173–194). Boca Raton, FL: CRC Press.

Hespanhol, L., & Tomitsch, M. (2015). Strategies for intuitive interaction in public urban spaces. *Interacting with Computers, 27*(3), 311–326.

Holyoak, K. J. (1991). Symbolic connectionism: Toward third-generation theories of expertise. In K. A. Ericsson & J. Smith (Eds.), *Toward a General Theory of Expertise* (pp. 301–335). Cambridge, UK: Cambridge University Press.

Hornecker, E. (2012). Beyond affordance: tangibles' hybrid nature. In *Proceedings of the Sixth International Conference on Tangible, Embedded and Embodied Interaction*: pp. 175–182: Kingston, ON, Canada ACM.

Hurtienne, J. (2009). Image schemas and design for intuitive use. PhD thesis, Technischen Universität Berlin, Germany.

Hurtienne, J., & Blessing, L. (2007). Design for intuitive use: Testing image schema theory for user interface design. Paper presented at the 16th International Conference on Engineering Design, Paris, France.

Hurtienne, J., & Israel, J. H. (2007). Image schemas and their metaphorical extensions: Intuitive patterns for tangible interaction. In Proceedings of the TEI'07, 1st International Conference on Tangible and Embedded Interaction (pp. 127–134). ACM, New York, NY.

Hurtienne, J., Klöckner, K., Diefenbach, S., Nass, C., & Maier, A. (2015). Designing with image schemas: Resolving the tension between innovation, inclusion and intuitive use. *Interacting with Computers, 27*, 235–255.

Ishii, H. (2008). *Tangible bits: Beyond pixels.* Paper presented at TEI'08, Second International Conference on Tangible and Embedded Interaction, Bonn, Germany (pp. 15–25).

Jacob, R. J., Girouard, A., Hirshfield, L. M., Horn, M. S., Shaer, O., Solovey, E. T., & Zigelbaum, J. (2008). Reality-based interaction: A framework for post-WIMP interfaces. Paper presented at the SIGCHI Conference on Human Factors in Computing Systems, Florence, Italy.

Jacoby, S., Gutwillig, G., Jacoby, D., Josman, N., Weiss, P. L., Koike, M., Sharlin, E. (2009). PlayCubes: Monitoring constructional ability in children using a tangible user interface and a playful virtual environment. Paper presented at the Virtual Rehabilitation International Conference, Haifa, Israel.

Jordà, S., Geiger, G., Alonso, M., & Kaltenbrunner, M. (2007). The reacTable: Exploring the synergy between live music performance and tabletop tangible interfaces. Paper presented at TEI'07, First International Conference on Tangible and Embedded Interaction, New York.

Lakoff, G. (1987). *Women, Fire and Dangerous Things: What Categories Reveal about the Mind.* Chicago: University of Chicago Press.

Lakoff, G., & Johnson, M. (1981). The metaphorical structure of the human conceptual system. In D. A. Norman (Ed.), *Perspectives on Cognitive Science* (pp. 193–206). Norwood, NJ: Ablex.

Langdon, P., Lewis, T., & Clarkson, J. (2007). The effects of prior experience on the use of consumer products. *Universal Access in the Information Society, 6*(2), 179–191.

LEGO. (2014). Lego.com.

Lewis, T., Langdon, P. M., & Clarkson, P. J. (2008). Prior experience of domestic microwave cooker interfaces: A user study. *Designing Inclusive Futures* (pp. 3–14). London: Springer.

Macaranas, A., Antle, A. N., & Riecke, B. E. (2015). Intuitive interaction: Balancing users' performance and satisfaction with natural user interfaces. *Interacting with Computers, 27*(3), 357–370.

McAran, D. (2018). Development of the technology acceptance intuitive interaction model. In A. Blackler (Ed.), *Intuitive Interaction: Research and Application* (pp. 129–150). Boca Raton, FL: CRC Press.

McEwan, M. (2017). The influence of naturally mapped control interfaces for video games on the player experience and intuitive interaction. PhD thesis, Queensland University of Technology, Brisbane, Australia.

McEwan, M., Blackler, A., Johnson, D., & Wyeth, P. (2014). Natural mapping and intuitive interaction in videogames. Paper presented at CHI PLAY '14, First ACM SIGCHI Annual Symposium on Computer–Human Interaction in Play, Toronto, Canada.

Merrill, D., Kalanithi, J., & Fitzgerald, B. (2011). Interactive play and learning system. US patent no. USD635190S.

Milgram, P., & Colquhoun, H. (1999). A taxonomy of real and virtual world display integration. *Mixed Reality: Merging Real and Virtual Worlds, 1,* 1–26.

Milgram, P., & Kishino, F. (1994). A taxonomy of mixed reality visual displays. *IEICE Transactions on Information and Systems, 77*(12), 1321–1329.

Mohan, G., Blackler, A. L., & Popovic, V. (2015). Using conceptual tool for intuitive interaction to design intuitive website for SME in India: A case study. Paper presented at IASDR 2015, "Interplay," Brisbane, Australia.

Mohs, C., Hurtienne, J., Israel, J. H., Naumann, A., Kindsmüller, M. C., Meyer, H. A., & Pohlmeyer, A. (2006). IUUI: Intuitive Use of User Interfaces. Paper presented at Usability Professionals 2006, Stuttgart, Germany

Muller-Tomfelde, C., & Fjeld, M. (2012). Tabletops: Interactive horizontal displays for ubiquitous computing. *Computer, 45*(2), 78–81.

Norman, D. (1993). *Things that Make Us Smart: Defending Human Attributes in the Age of the Machine.* Reading, MA: Addison-Wesley.

Norman, D. (1999). Affordance, conventions, and design. *Interactions, 6*(3), 38–43.

Norman, D. (2010). Natural user interfaces are not natural. *Interactions, 17*(3), 6–10.

Norman, D. (2013). *The Design of Everyday Things, revised and expanded edition.* New York: Basic Books.

O'Brien, M. A., Rogers, W. A., & Fisk, A. D. (2008). Developing a framework for intuitive human–computer interaction. Paper presented at the 52nd Annual Meeting of the Human Factors and Ergonomics Society, New York.

Skalski, P., Tamborini, R., Shelton, A., Buncher, M., & Lindmark, P. (2011). Mapping the road to fun: Natural video game controllers, presence, and game enjoyment. *New Media & Society, 13*(2), 224–242.

Smith, D., Irby, C., Kimball, R., & Verplank, B. (1982). Designing the star user interface. *Byte, 7*(4), 242–282.

Still, J. D., & Dark, V. J. (2010). Examining working memory load and congruency effects on affordances and conventions. *International Journal of Human–Computer Studies, 68,* 561–571.

Still, J. D., & Still, M. L. (2018). Cognitively describing intuitive interactions. In A. Blackler (Ed.), *Intuitive Interaction: Research and Application* (pp. 41–62). Boca Raton, FL: CRC Press.

Still, J. D., Still, M. L., & Grgic, J. (2015). Designing intuitive interactions: Exploring performance and reflection measures. *Interacting with Computers, 27*(3), 271–286.

Tangible Play. (2014). Osmo: https://www.playosmo.com/en/.

Tretter, S., Diefenbach, S., & Ullrich, D. (2018). Intuitive interaction from an experien-
tial perspective: The intuitivity illusion and other phenomena. In A. Blackler (Ed.),
Intuitive Interaction: Research and Application (pp. 151–170). Boca Raton, FL: CRC
Press.

Vallgårda, A. (2014). Giving form to computational things: Developing a practice of interaction design. *Personal and Ubiquitous Computing, 18*(3), 577–592.

Vera, A. H., & Simon, H. A. (1993). Situated action: A symbolic interpretation. *Cognitive
Science, 17,* 7–48.

3 Cognitively Describing Intuitive Interactions

Mary L. Still and Jeremiah D. Still

CONTENTS

3.1   INTRODUCTION

Interaction designers create interfaces with the hope of facilitating the user's ability to complete a task (Simon, 1969). Also, it is often the goal to ensure that the interaction with the interface stays at the periphery of attention (Matthies, Muller, Anthes, & Kranzlmuller, 2013). This goal is not always attained. Many of the interfaces we regularly interact with require focused attention. Those interfaces that do not require extra cognitive effort are often described as being intuitive, implying that the interactions take little effort to complete. Historically, though, there has been little agreement about what makes an experience intuitive or how that experience ought to be measured. Left unconstrained, this seemingly important design concept could be rendered theoretically useless (see Allen & Buie, 2002).

After nearly two decades, there has been progress. Many now agree that intuitive interactions are characterized by the effortless application of previous knowledge (e.g., Allen & Buie, 2002; Hurtienne & Israel, 2007; Naumann et al., 2007; Naumann &

Hurtienne, 2010; Spool, 2005; Wippish, 1994). And although there is no consensus about the best way to measure the intuitiveness of a design, there are a set of common instruments: the Technology Familiarity Questionnaire (Blackler, Popovic, & Mahar, 2003), the Questionnaire for the Subjective Consequences of Intuitive Use (QUESI; Naumann & Hurtienne, 2010), and INTUI (Ullrich & Diefenbach, 2010a, 2010b). (For a full review of methods for intuitive interaction, see Chapter 4; Blackler, Popovic, & Desai, 2018). Even with this progress, there remains some ambiguity about the processes and representations that support intuitive interactions. Without a clear understanding of how intuition operates, it is hard to predict what interactions will be intuitive. This is particularly problematic when designing novel interactions—for example, for new product types or technologies (e.g., Fischer, Itoh, & Inagaki, 2015).

In this chapter, we consider how understanding a set of basic cognitive processes—implicit and incidental learning, knowledge representation, familiarity, transfer, and automaticity—can help explain how an interaction can be intuitive or become intuitive over time. Critical to this discussion is the process by which engaging in similar interactions can come to support the intuitive use of another interaction. We suggest that as this compilation of experiences is stored in long-term memory, it facilitates interactions that fall on a gradation of intuitiveness (the Intuitive Interaction Hierarchy, depicted in Figure 3.1). In addition to determining the intuitiveness of an interaction, these graded representations constrain what information the user has metacognitive access to. The more intuitive an interaction is to use, the less likely the user will be able to accurately describe why the interaction is intuitive. We examine the empirical data relevant to these limitations in knowledge elicitation.

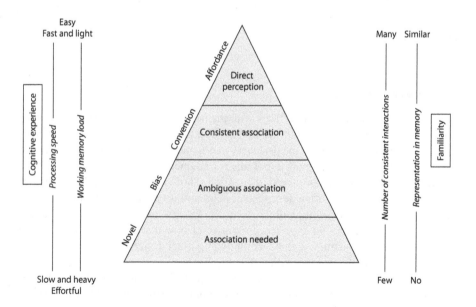

FIGURE 3.1   Intuitive Interaction Hierarchy.

3.2 COGNITIVE PROCESSES THAT SUPPORT INTUITIVE INTERACTIONS

We believe the cognitive science literature can provide the theoretical foundation for understanding what processes drive intuition. Even when key processes are identified, there can be difficulty translating cognitive research into actionable recommendations, in part due to the differences in scope between cognitive studies and design studies. Cognitive experiments are often designed to test the mechanisms underlying intuition by creating experimental conditions to differentiate between those proposed mechanisms. This stands in contrast to most design studies that are primarily concerned with assessing design use in a specific context. Cognitive researchers typically use speed and accuracy to examine intuition. By comparison, researchers examining design for intuitive interaction may use speed and accuracy measures but also rely heavily on participants' subjective experiences and/or researchers' coding of video data. Another difference lies in the definition, with cognitive scientists generally describing intuition as an automatic process that is based on implicit knowledge and results in a signal or feeling that is used to make a decision (e.g., Betsch, C. 2008; Betsch, T., 2008; Epstein, 2010; Hodgkinson, Langan-Fox, & Sadler-Smith, 2008; Reber, 1989; Topolinski & Strack, 2009). This distinction is potentially important because cognitive research emphasizes the implicit nature of the knowledge supporting intuition, while the design literature does not. The concept of implicit learning has been discussed as a possible route to intuitive interaction in the design literature (Blackler, Popovic, & Mahar, 2010), but this literature does not explicitly specify how previous knowledge supporting intuitive interactions is obtained.

Even though there are some differences between these fields in the way intuition has been investigated, a better understanding of cognitive principles can lead to a better understanding of how intuitive interactions emerge (see Hodgkinson et al., 2008 for a similar assertion). Still, Still, and Grgic (2015) suggest that an interaction will be intuitive if the mental representation of the current interaction is similar to one already stored in memory and the action associated with the stored representation is automatically applied to the current situation. Regarding cognitive phenomena, this means that researchers must consider the nature of implicit learning and memory, the mental representations that support familiarity, the mechanisms associated with transfer, and the conditions necessary to support automatic processing.

3.2.1 IMPLICIT LEARNING

Interactions do not occur in a vacuum. Instead, users' interaction experiences are always shaped by their interaction histories. When it comes to learning how to use an interaction, a series of discrete experiences may come to mind. For instance, an individual might remember taking a typing class or using a training program to help them learn how to use a QWERTY keyboard more effectively. While these explicit experiences, where an individual intentionally invests time and effort in the learning process, contribute to the knowledge that is stored in memory, learning can occur in the absence of such concerted effort. Incidental learning occurs when an individual learns how to complete a task or learns the rules that govern an interaction by indirect exposure to the stimuli that follow those rules.

This type of learning can be more pronounced when an individual learns the rules without any insight into what the rules might be. There are many examples of this implicit learning in the cognitive literature. One of the most cited examples involves language learning (e.g., artificial grammar). In this paradigm, participants are exposed to what appear to be nonsense strings; these strings, though, are based on a set of rules known to the researcher. Critically, the initial exposure to the stimuli need not focus on the sequencing at all; Manza and Bornstein (1995) demonstrated the effect just by having participants rate how much they like or dislike a letter string. With repeated exposure to this artificial grammar, participants can accurately identify "legal" sequences (e.g., Reber, 1967). In other words, the statistical properties of a set of stimuli can be extracted from these experiences. Further, even though the rules of the artificial grammar can be applied under new conditions, participants are typically unable to verbalize what the rules are (e.g., Reber, 1967, 1989). This type of statistical learning has been demonstrated in a variety of situations, including language learning, categorization, sequence/probability learning, production/economic systems, and location prediction (e.g., Reber, 1967; Fried & Holyoak, 1984; Cleeremans & McClelland, 1991; Broadbent & Aston, 1978; and Lewicki, Hill, & Bizot, 1988; respectively). While it is true that explicit instruction on word usage and grammar are helpful, much of the learning of these rules could be accomplished implicitly. Interestingly, in his review of the implicit learning literature, Reber (1989) suggested that efforts to learn rules might even interfere with implicit learning. In support of this claim, Cleeremans and McClelland (1991) found no evidence of a significant contribution of explicit knowledge to the performance of a sequence detection task used to simulate grammar learning. From a design perspective, at a minimum, these findings suggest that learning can occur in a variety of circumstances and those circumstances may be difficult to anticipate. Therefore, a seemingly arbitrary user experience decision may not be arbitrary in the broader context of learned interactions.

3.2.2 MENTAL REPRESENTATION AND FAMILIARITY

Representations of these past interactions are stored in long-term memory and are accessed and applied when similar interactions are encountered. As early as 1994, Raskin suggested that an intuitive interaction is not much more than a familiar interaction. It is typically assumed that to apply previous knowledge to a new situation, there must be some process that compares previous experiences with the current experience (Mandler, 1980). The outcome of that process is often referred to as a *familiarity signal*. The familiarity signal can be vague—participants may not know why something feels familiar—but it may be used to support other decisions such as whether or not an unidentifiable item is a real object, whether or not an unidentified face or scene is famous, and whether or not a stimulus is threatening (see Cleary, Ryals, & Nomi, 2013). With repeated exposure, the familiarity of a stimulus increases; the features that represent the stimulus become more strongly associated in memory (e.g., Atkinson & Juola, 1974; Mandler, 1980). A closely related concept is that of *directness*. Directness is a measured distance between the user's mental representation (i.e., stored representations of previous experiences) and a system's representation (i.e., the representation activated by the current interaction). This distance is known as a *knowledge gap*. According to Spool (2005), an intuitive experience is created by reducing the difference between users' current knowledge and the knowledge required to use the system effectively. Designers need to first capture what the users' current knowledge is and contrast it with the required system knowledge to identify the gap. Spool highlights this process as the "biggest challenge" in creating intuitive experiences.

In order to predict *a priori* what interactions will be familiar, some understanding of the mental representations supporting the interaction is required. But, as Cleary, Langley, and Seiler (2004, p. 903) note, "little is known about the types of features that may actually be present in human memory traces." Nevertheless, the cognitive literature does provide some speculation about features stored in memory. The features of interest vary depending on the area of research but include letter features, letters, phonemes, morphemes, syllables, words, grammar, concepts, category membership, color, line orientation, and schemas to name a few. When it comes to design, it is unclear what the underlying mental representations would be. They may be features within a class of items (e.g., cameras and phones) or even features that are encountered across interaction categories (e.g., remote controls, turn signals, escalators; Blackler, Desai, McEwan, Popovic, & Diefenbach, 2018; Fischer et al., 2015; Still et al. 2015).

There are likely representations of the actions associated with particular features as well. Hurtienne and Israel (2007) discuss image schemas as a way to conceptualize the representations that would be applicable in user interface design. Image schemas are rich representations as they include multimodal representations and sensorimotor experiences. Images schemas include those pertaining to space (e.g., relative movement: up–down, left–right; path), containment (e.g., containers: things that go together), force (e.g., obligations, emotions), and attribute (e.g., recurring properties of a specific entity).

3.2.3 KNOWLEDGE TRANSFER

When sufficient overlap between a stored representation and the representation of the current situation occurs, knowledge about the previous experience can be transferred, or applied, to the current situation. This transfer of knowledge can occur automatically without intention on the part of the user. When previous experience aligns with the current interaction, positive transfer can occur and the interaction may seem intuitive. When past experience provides an incorrect mapping for the current interaction, negative transfer can occur (Besnard & Cacitti, 2005). Both types of transfer are pertinent in the design world. According to Still and Dark (2013), disconnects between the interface design and the user's previous knowledge may lead to unsuccessful interactions. Additionally, in some instances, the negative transfer can result in phantom affordances that cause frustration. For example, imagine a user visits a web page and clicks on an underlined word—nothing happens. From past experience, the user expects underlining to indicate the presence of a hyperlink, but in this case the designer used underlining to indicate that the word was a heading. This violation of previous use cases produces a phantom affordance. The user believes the underlined word is clickable, but in this context the clicking action is not available. This incorrect mapping of visual element to expectation leads to an unintuitive interaction.

In addition to causing frustration and increasing the time required to learn how to use an interface, negative transfer can lead to increased cognitive load during the interaction. Imagine that you are an expert QWERTY keyboard typist and you have purchased a device with a different keyboard configuration. You attempt to type an "a" and the device returns a "k" instead. Your past QWERTY expertise now hinders your performance. You find yourself spending limited cognitive resources focusing on letter selection, rather than focusing on higher-level tasks (e.g., effectively communicating your ideas). Although it might seem unlikely that a keyboard would be remapped, there are interfaces where the letters are arranged in alphabetical order (e.g., GPS, gaming systems).

In contrast to the difficulties faced with negative transfer, interactions can seem intuitive and be learned quickly (Keppel & Underwood, 1962) when there is positive transfer. Consider another scenario in which you encounter a QWERTY keyboard layout in a new context: on your cellular phone. This interaction requires you to use only your thumbs to type, which can be a challenging task. Despite this difference, you can quickly and effortlessly select letters because you do not have to think about where the letters are located. Although it may be cumbersome to type with your thumbs, it is likely that the interaction itself will seem intuitive because of the positive transfer of knowledge.

3.2.4 AUTOMATICITY

Interaction consistency is the key to producing automatization, which leads to a lighter working memory load. Controlled processing of interface interactions results in a user's awareness of the actions (Posner & Snyder, 1975). However, automatically processing interactions does not allow for awareness of the underlying processing that produces the response. Typically, technology interactions are a mixture of both types of information processing (Shiffrin & Schneider, 1977). For instance, many driving interactions become automated, such as using a turn signal before taking a left or right turn. The underlying action mapping is consistent. This is similar to using a steering wheel or operating a manual transmission. These automatic processes carry little working memory load. These mappings are stored in memory within a schema for expert drivers (Norman & Shallice, 1986). However, certain driving tasks will always remain under controlled processing because elements of the interaction are dynamic and, therefore, do not have an established schema. For instance, although navigating home via a familiar route can be relatively automated, finding an alternate route to avoid a traffic accident would not be automated. The controlled processing associated with way-finding also carries a heavy working memory load.

It seems that there are also many cases in which an unintuitive design can become intuitive with enough experience (see Spool, 2005; training program recommendation). Some interactions simply will not be intuitive on first use, but that does not mean they never will be. The QWERTY keyboard is a prime example. The first interaction with a QWERTY keyboard is unlikely to be characterized as intuitive. However, after years of experience with the QWERTY keyboard, it may seem intuitive, particularly in comparison with other key configurations. Hurtienne and Israel (2007) make a similar assertion in suggesting that frequent interaction with a stimulus increases the likelihood that previous knowledge can be applied without awareness.

Depending on their complexity, arbitrary mappings between design elements and function require varying amounts of cognitive effort to learn. A scrollbar, for instance, has little interaction complexity and requires few attentional resources to learn. In contrast, learning the spatial locations of each key on a keyboard requires a significant amount of attentional resources. These attentional costs, though, are not necessarily static. When the mappings between design and function are consistent, over time the amount of working memory required to complete the task decreases, thereby reducing cognitive load. This occurs because every interaction provides an opportunity to learn consistencies in the interaction mapping and those consistencies are stored in long-term memory. Therefore, when the interaction is experienced again, the user already has the interaction mapping in memory and does not have to rely on limited working memory resources to perform the correct response. In this way, knowledge structures are relied on to free cognitive resources. The simpler and more consistent the interaction, the faster this process can take place. We propose that this direct connection between the interaction and long-term memory is a fundamental component in the intuitive experience.

3.3   USING COGNITIVE PRINCIPLES TO INFORM DESIGN

To simultaneously consider the influence of these cognitive processes on interactions, we propose the Intuitive Interaction Hierarchy (Figure 3.1). At the core of the hierarchy are cognitive processes thought to underlie intuitive interactions. Each process is represented by a continuum. This illustrates the assumption that the "location" of any interaction in the hierarchy is malleable given changes in experience, competing interactions, and workload. Included in the hierarchy is the consideration that some interactions benefit from experience, whereas others are intuitive simply because their usage is highly constrained or is immediately obvious based on perceptual processing (Still et al., 2015). For instance, the top button of a vertically orientated pair affords moving up within a menu.

As depicted in Figure 3.1, the pinnacle of the Intuitive Interaction Hierarchy is an interaction that can be completed effortlessly. The mechanism that supports this effortless use is automatization. Automatization is only possible when the interaction between an interface element and its function is consistent and that association is stored in memory. In our Intuitive Interaction Hierarchy, the *direct perception* (Gibson, 1979) level of intuitiveness is achieved through these consistent associations and may even be supported by interaction experiences across a variety of circumstances (e.g., using a knob to increase/decrease the volume on a stereo, the temperature on a heater, or fan speed). The advantage to using this type of interaction is that processing requires few cognitive resources, allowing the bottleneck of awareness to be circumvented.

When affordances cannot be used, designers have to make decisions about how to map interactions that may otherwise seem arbitrary. Under these circumstances, a convention, bias, or even a novel interaction mapping may be employed. In each case, users are required to learn, in one form or another, how the interaction should unfold, but the intuitiveness of these interactions will depend on previously existing mental representations of the task. Conventions are the most regularized examples of arbitrary interactions and we describe them as having *consistent associations*. As the name suggests, these interactions benefit from clear mappings; they may be characterized by fewer interaction experiences or by relatively inconsistent interaction experiences compared with those that are directly perceived. Interactions that are even less practiced or more irregular are described as having *ambiguous associations*. We classify these as biased interactions because they are used with some consistency (above chance level), but the interactions may lack clear mappings as a result of negative transfer or may be relatively new interactions for the individual such that mental representations of the interaction are still developing. Finally, if an interaction has no corresponding mental representation, it is clearly in need of memory association. These novel interactions may be completely new to the individual and still require additional analysis to form mappings in long-term memory. In these cases, the user may have no expectation for how the interaction should perform; as he or she gains experience with the interaction, he or she can test the hypotheses made about the interaction, thereby forming some coherent representation of the task. The hierarchy illustrates how increases in intuitiveness correspond with faster processing and lighter working memory load.

The scrollbar provides a good example of a design decision that can be "located" at various levels in the hierarchy, depending on user experience. The function of a scrollbar is to move the content in the central frame either up or down. Should the interface be designed so that the central content moves up when the bar moves up or should the content move down? This original decision of the button-to-action mapping is arbitrary. However, after gaining experience with one particular scrollbar mapping, the interaction becomes less arbitrary to the user. If, for example, users have extensive experience moving the scrollbar down to move content up the page, they would not find the opposite mapping (moving the scrollbar down to move content down) to be intuitive, even though they have had experience with scrollbars and up–down image schemas in the past.

3.4 CONSIDERING INTUITIVE INTERACTION LEVEL WHEN SELECTING ASSESSMENT METHODS

Interactions vary along a continuum in terms of their potential to be intuitive. Some interactions are unlikely to become intuitive because they are too complex or the users' experience with the mappings is too varied. Other interactions have the potential to become intuitive interactions with enough experience and consistency. Finally, some interactions are intuitive simply because their mappings are "natural." One of our goals in mapping affordances, conventions, and biases onto intuitive interactions was to draw attention to the fact that although some interactions are intuitive on their first novel use, other currently non-intuitive interactions have the potential to become intuitive with adequate and consistent experiences (e.g., Hurtienne & Israel, 2007; Still et al., 2015).

3.4.1 Affordances

Gibson (1979) introduced the concept of affordances, but Norman (1988) introduced the human–computer interaction community to it. According to Norman (1988, p. 9), "affordance refers to the perceived and actual properties of the thing, primarily those fundamental properties that determine just how the thing could possibly be used." For example, plates on doors are for pushing and kitchen cabinet knobs are for pulling. According to You and Chen (2007), affordances act as intuitive actions that support the user–product interaction. Further, Blackler (2008) states, "The concept of affordance suggests a route to intuitive use" (p. 92). She highlights that through direct perception, affordances ought to support the initial interactions with a novel interface. One could consider affordances to be naturally intuitive. This intuitive use could come about from a user's familiarity with how to use the object (e.g., Raskin, 1994).

3.4.2 Conventions

In some circumstances, it is not possible to use affordances in an interaction. Instead, an "arbitrary" mapping might be used. Even when a mapping is considered arbitrary, users tend to adopt a preferred method of interaction. When a particular preference is observed across a group of individuals, it is referred to as a *population stereotype* (Bergum & Bergum, 1981). Bergum and Bergum suggest that preference levels greater than chance, even if they are only slightly above chance level, may be useful from a design perspective. Given that different levels of automaticity and familiarity may support interactions, we believe it is useful to consider different levels of nonaffordance interactions. We refer to those interactions that are most regularized as conventions and those that are less regularized as biases. Critically, we propose that the differences between conventions and biases emerge based on the interaction history of the individual. This focus on individual interaction history and the resultant mental representations presents a slightly different approach to examining intuitive interactions than the traditional use of population stereotypes.

Skilled use of a keyboard provides an example of interactions that are based on conventions. There is no affordance guiding the position of the "F" key on a QWERTY keyboard, but a decision had to be made about "F" key placement. This does not mean that use of the "F" key will never be intuitive. Once the user learns a consistent, arbitrary mapping, that mapping can be acted on rapidly with little demand on cognitive resources. For instance, pressing the "F" key on a keyboard is natural for expert typists, even when they cannot see the keyboard. In fact, expert typists can press the "F" key without "thinking" about its placement (see Rumelhart & Norman, 1982). This type of learned interaction has been referred to as a *convention* (Norman, 1999). According to Still and Dark (2008), affordances and conventions can be behaviorally indistinguishable. This finding demonstrates that through enough regular interactions, an entirely novel interaction based on a convention may appear to be as intuitive as an affordance. In essence, learning occurs through these regular interactions and removes ambiguity from originally arbitrary interactions (Logan, 2002). Raskin (1994) suggested that the "intuition" experienced from a convention also reflects familiarity with the object and its uses. These perspectives are congruent with a cognitive approach (Still & Dark, 2013). From a cognitive perspective, every interaction experience is integrated into memory. If the interaction is novel and does not resemble other interactions, a new representation for the interaction will be established. If the interaction is familiar, then each new experience with the interaction will contribute to the existing representation. Over time, representations of regularized interactions can be accessed with little or no cognitive effort. An example of a convention appears in Figure 3.2b. In this device, two vertically oriented buttons labeled with double arrows are used to move forward (right) and backward (left) through phone messages. Based on data from Still and Dark (2008), a convention exists for using the top button to move right, thus the phone interaction is congruent with the convention (in addition, the icons used in this interface have been standardized to ISO/IEC standard 18035; ISO/IEC, 2003).

3.4.3 BIASES

Unlike conventions, some non-affordance interactions are less intuitive and are defined by the inconsistent behaviors associated with them. From a behavioral perspective, a biased interaction is one that is used above chance level (e.g., when given two options for completing an interaction, one of the options is selected more than 50% of the time) but is still below the level of conventional use. Whereas a convention reflects a standardization of an interaction—often resulting in automatic or skilled responses—a bias reflects some lack of familiarity or inconsistent interaction

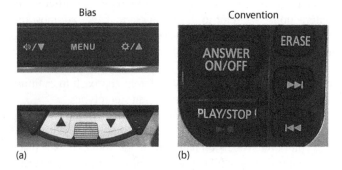

FIGURE 3.2 Real-world examples of interactions supported by biased and conventional representations. (a) An interaction that uses a mapping consistent with the biased response (top example from a monitor) and an interaction with the opposite button-to-action mapping (bottom example from a sit/stand desk). (b) An interaction (answering machine) that uses a mapping consistent with the conventional use for arrow directionality.

experiences. Because of this, a bias is associated with greater distance between the interaction and the stored representation in long-term memory compared with conventions. We suggest that biases can even emerge from inconsistent interactions across interfaces. An example of a bias appears in Figure 3.2a. Both devices use two horizontally oriented buttons; the top image shows a computer monitor where the left button is used to decrease volume and brightness, while the right button is used to increase volume and brightness. The bottom image shows the controls for a sit/stand desk where the left button is used to increase desk height, while the right button is used to decrease desk height. In these examples, there are two competing interaction mappings; the button on the right represents increase in one instance and decrease in the other. This ambiguity prevents the establishment of a convention in long-term memory. Based on data from Still and Dark (2008), the biased interaction is to move up using the right button and down using the left button. Therefore, the mapping for the monitor buttons should be preferred and should be slightly more intuitive than the mapping for the sit/stand desk.

The challenge in identifying these cases is that even interactions with other devices are included in a user's past experience. For instance, while an individual might interact with a variety of remote controls that use the top button to move right, the user may encounter the same mapping when driving a car. In the United States, a vehicle's turn signal is moved "up" to indicate an upcoming right turn. This implicit interaction knowledge can then be applied to other similar tasks. Therefore, when users are presented with a consciously ambiguous decision, they feel the top button ought to be mapped to the right. This feeling reflects the influence of familiarity. This example highlights the difference between conscious and unconscious influences. A designer might easily believe an interface has a truly arbitrary mapping, but unconscious familiarity signals clearly influence users. The designer's ability to perceive the interface in a similar way to the user depends on his or her amount of shared interaction experiences. One challenge to this is simply that the designer may not share interaction experiences with the target user; perhaps the designer has more technological expertise, for example. A second challenge is that while the designer is creating a new interaction experience, his or her mental representation of the task will be more practiced and sophisticated than that of a naïve user.

3.4.4 IMPLICATIONS OF PREVIOUS EXPERIMENTS

To better understand how designers might capture accurate user interaction knowledge, Still, Still, and Grgic (2015) used simple interactions that varied in intuitiveness while also using two knowledge elicitation methods. The interaction consisted of three different two-button configurations that had to be used to navigate left, right, up, and down. The buttons were located side by side with a horizontal orientation, one above the other with a vertical orientation, or they were diagonal from one another. These specific interactions were used because they represent bias (using the horizontal buttons to move up and down), convention (using the vertical buttons to move left and right), and affordance conditions (using horizontal buttons to move left and right, vertical buttons to move up and down, and using the diagonal buttons to move in any direction). Data from Still and Dark (2008) were used to identify the bias (e.g., right button for moving up) and convention (e.g., top button for moving right) interactions. Importantly, participants acted on each button configuration moving in all four directions—to provide an objective measure of participants' interaction knowledge—and they assigned button-to-action mapping for each button configuration during a reflective task—a measure of participants' subjective, or explicit, interaction knowledge. The primary result of interest was the fact that the effectiveness of the knowledge elicitation tasks varied along with the level of intuitiveness of the interaction. For affordance interactions, both measures yielded the same results. For conventions, individuals completing the performance task first followed convention and were likely to provide the same interaction mapping for the reflective task (92% of participants reported that they would use the convention) as they used in the performance task (89% used the convention). In contrast, when participants completed the reflective task first, they were less likely to select the conventional mapping (only 27% of participants reported that they would use the convention), but interestingly, these same individuals were likely to use conventional mappings when completing the performance task (59% used the convention). A similar pattern of results was obtained for bias interactions although the order effects were less pronounced. The results of this experiment are important to consider because they show that when a researcher is attempting to discover what interactions might be intuitive, the way that knowledge is elicited can have a significant impact on the results. In this particular situation, if Still et al. had only used a reflective task to determine which button-to-action mapping to use for the vertically oriented buttons, they would have mapped the buttons in the opposite direction (top button for moving left) of the existing convention.

We propose that non-affordance interactions that have consistent associations (e.g., conventions) may be supported by implicit processes that are not available for conscious report. Thus, when asked how the interaction should be designed, the participant can provide an answer, but that answer may not reflect their actual experience. Non-affordance interactions that have more ambiguous associations may not be consistent enough or practiced enough to trigger automatic processes. Because of this, participants may still be using deliberate processes to complete the interaction, and they should at least have access to those intentional acts. Biased interactions are still supported by implicit knowledge but not to the extent of conventional interactions. Previous research corroborates this interpretation of the results; Ericsson and Simon (1993) suggest that knowledge elicitation is most effective when the measure used to elicit the knowledge is supported by the same processes as the knowledge itself. There is substantial evidence that dissociations can exist between participants' actions and participants' subjective assessments of those actions (e.g., Andre & Wickens, 1995; Bailey, 1993; Reber, 1967).

3.5   CONSIDERING THE PHENOMENA MEASURED WHEN SELECTING ASSESSMENT METHODS

Intuitive interactions should require little effort and they ought to be based on familiar interactions or interface elements. Many designers would also contend that the interaction should be subjectively experienced as being easy to use. Considering these requirements, it seems that many of these characteristics could be captured with common usability measures. Therefore, we designed a pilot study to examine the effectiveness of traditional usability measures at indexing intuitive interactions in comparison with instruments designed specifically to measure intuitive interactions. For this investigation, the NASA Task Load Index (NASA-TLX; Hart & Staveland, 1988) and the System Usability Scale (SUS; Brooke, 1996) were the usability measures; the QUESI (Naumann & Hurtienne, 2010) and a questionnaire inspired by Blackler et al.'s (2003) Technology Familiarity Questionnaire were used to index intuitiveness. Response time, error rate, and familiarity ratings were also collected to ensure that the interactions we selected for the study varied in difficulty. Twelve students participated in this repeated-measures design. Log-in task order was counterbalanced across participants. Due to the limited size, a Latin square was used to counterbalance the order in which participants received the four intuitiveness measures; this resulted in each of the four pseudorandom assessment orders being represented three times across the data set.

The interactions used in this experiment were the three authentication log-in schemes represented in Figure 3.3. One log-in scheme was intended to be very familiar, while the other two were expected to be unfamiliar, with one being harder to use than the other. The goal was to have interactions that would represent varying levels of intuitiveness. The familiar interaction, shown in Panel A of Figure 3.3, was a PIN entry system. Participants were assigned four numbers and entered those numbers in sequential order by selecting them with a mouse on a screen-displayed number pad.

The relatively easy but unfamiliar authentication scheme in Panel B used the *convex hull click* (CHC) method (Wiedenbeck, Waters, Sobrado, & Birget, 2006). In this method, participants are given three images as their "passcode," then they are presented with an array of images on the screen. To authenticate, the participant has to visually locate their images on the computer screen, mentally draw a line connecting those images, and then click somewhere inside the body of the imaginary shape formed by the connected line. In Panel B of Figure 3.3, the *T*s represent the passcode images, *D*s are distractor images, and the gray area shows where participants could click to log in. The relatively difficult and unfamiliar authentication scheme used (Panel C) was the *Where you see is what you enter* (WYSWYE) technique (Khot, Kumaraguru, & Srinathan, 2012). In this method, participants are given a set of images to remember, then are shown a grid containing images in each position; one row and one column in the grid will not have any of the passcode images. Participants are instructed to find the row and column that do not have any passcode images (second column and fifth row in the example, as signified by the dotted borders in Panel C of Figure 3.3), then imagine that the row and column have been removed to create a new grid. The four passcode items should still "appear" in the participants' mental grids. Then, participants are shown a blank grid and have to select the grid locations where their passcode images appeared in the imagined, transformed grid. The grid on the right in Figure 3.3 shows which locations the participant should select. These locations correspond to the locations of the passcode images (*T*s) once the "extra" column and row have been removed. The 12 participants in the study all completed 10 trials using each log-in method; they were assigned a new passcode

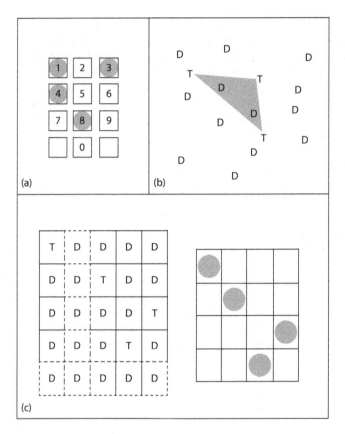

FIGURE 3.3 Authentication schemes: (a) PIN entry, (b) convex hull click, (c) where you see is what you enter.

every time. After interacting with a log-in method, participants completed familiarity and intuitiveness measures.

Importantly, the three log-in techniques did reflect the predicted different levels of intuitive use. Participants reported greater familiarity with the PIN entry method, were faster at using the method, and made very few errors (see Table 3.1). Participants were less familiar with the CHC and WYSWYE methods and were both slower and more error prone using those techniques, with the WYSWYE method being the most difficult.

After interacting with each log-in task, participants completed a battery of scales intended to assess their subjective experiences: NASA-TLX, SUS, QUESI, and an experience questionnaire. The NASA-TLX measures mental demand, physical demand, temporal demand, performance quality, effort level, and frustration level (Hart & Staveland, 1988). We used the Raw TLX (RTLX) instead of the weighted score, a practice that has become more common (see Hart, 2006). The SUS (Brooke, 1996) measures users' subjective experiences using a system or product. Items include "I thought the system was easy to use," "I found the various functions in this system were well integrated," and "I would imagine that most people would learn

TABLE 3.1

Condition Means for Measures Indexing Interaction Difficulty (Standard Deviation in Parentheses)

	Familiarity	Log-in Time (ms)	Proportion Log-in Failures
PIN	4.50 (0.86)1,2	3579 (423)3	0.03 (0.05)5,6
Convex hull click	1.93 (0.75)1	12417 (5597)4	0.32 (0.27)5
WYSWYE	2.22 (0.92)2	38936 (18263)3,4	0.41 (0.26)6

Note: Variability in the log-in time for the CHC and WYSWYC methods comes in part from the high error rate; only correct interactions were included in the log-in time measure. Familiarity was measured on a scale of 1 (low) to 5 (high). Repeated-measures ANOVAs were used to analyze the data; post hoc analyses were only conducted when a significant main effect was obtained. Comparisons denoted by 1, 2, 3, 4, 5, and 6 were significantly different ($p < .05$ in post hoc analyses with Bonferroni corrections).

to use this system very quickly." Although the scale is most often used as a unitary measure, Lewis and Sauro (2009) identified two factors in the survey questions that index perceived ease of use and perceived effort (usability and learnability), two critical qualities of intuitive design. The QUESI measures five constructs: mental workload, goal achievement, the effort of learning, familiarity, and perceived error (Naumann & Hurtienne, 2010). Our authentication experience questionnaire asked participants to report how often (if ever) they used the log-in technique and how much additional experience they believed they needed to effectively use the log-in technique, and asked them to estimate how many different authentication methods they have used that include critical features or processes associated with that technique. For example, to assess the CHC method, we asked participants about their experiences memorizing images, memorizing image location, using invisible or imagined shapes, and using specific locations on a screen to complete an authentication task. This questionnaire was inspired by Blackler et al.'s (2003) Technology Familiarity Questionnaire; in our questionnaire, we identified essential elements of the interaction and asked participants to report their experience with those features. We depart from Blackler et al. in one key aspect: the Technology Familiarity Questionnaires presented specific products to the participants and asked them to report their product and associated feature usage, whereas we did not ask about specific products. Because people do not have broad knowledge about authentication methods—and especially do not know the names of those methods—we asked them about their experiences with specific aspects of the interaction, such as remembering a number sequence or creating a mental image; this was done in addition to asking specifically about their interaction histories with the authentication method. Three versions of our authentication experience survey were created, one for each interaction. Because there was the potential for carryover effects between the measures, scale order was counterbalanced across participants.

The subjective measures show that all four were sensitive to the differences between the interactions (see Table 3.2). The QUESI was the most sensitive, showing

TABLE 3.2

Condition Means for Measures Indexing Intuitive Interaction

	Raw TLX	SUS	QUESI	Experience
PIN	21.3 (9.4)1	70.6 (7.1)	4.78 (0.31)4,5	75.4 (16.8)7,8
Convex Hull Click	31.7 (16.9)2	72.7 (6.4)3	3.92 (0.80)4,6	25.2 (13.5)7
WYSWYE	50.2 (21.0)1,2	65.0 (9.4)3	2.44 (0.45)5,6	15.6 (9.8)8

Note: Standard deviation in parentheses. Repeated-measures ANOVAs were used to analyze the data; post hoc analyses were only conducted when a significant main effect was obtained. Comparisons denoted by 1, 2, 3, 4, 5, 6, 7, and 8 were significantly different ($p < .05$ in post hoc analyses with Bonferroni corrections).

significant differences between all three interactions; the authentication experience questionnaire and the NASA-TLX exhibited qualitatively similar sensitivity (although an examination of the data show that the measures do not make the same distinctions between interaction type); and the SUS was the least sensitive of the measures, only indexing a difference between the CHC and WYSWYE interactions.

3.6 DISCUSSION

Based on the examination of these specific measures, it appears that the NASA-TLX provides a relatively sensitive index of intuitiveness, but the SUS may not. To determine if the NASA-TLX could serve as a reasonable proxy for an intuitive interaction measure, it is helpful to directly compare the NASA-TLX with the authentication experience questionnaire. Clearly, the NASA-TLX is focused on workload and cognitive effort. In comparison, the authentication experience questionnaire provides only an indirect measure of workload (How much more experience would you need to be proficient?) and focuses primarily on indexing familiarity with the interaction elements. The different foci of the measures are clear in the data, as the NASA-TLX differentiated between log-in methods that were cognitively easier (i.e., fewer mental transformations) and more difficult (i.e., WYSWYE). The authentication experience questionnaire, in contrast, differentiated between log-in methods that were more familiar and less familiar (i.e., CHC and WYSWYE). It is possible that with a larger sample these two measures would have provided further differentiation between the three log-in techniques, but even with the current data set, the strengths of each measure are apparent.

Overall, the QUESI was the most sensitive measure, but with questions relating to mental workload, goal achievement, the effort of learning, familiarity, and perceived error—nearly every construct covered in the other measures—perhaps this result is not too surprising. QUESI items assessing mental workload, goal achievement, and perceived error parallel NASA-TLX items for mental workload, performance, effort, and, potentially, frustration. QUESI items assessing familiarity may parallel some aspects of familiarity collected via the authentication experience questionnaire (even if the approach is different). With these similarities in mind, one might wonder if the

NASA-TLX could simply be used in conjunction with some reliable index of familiarity to measure intuitive interaction.

Although the SUS is a mainstay in usability testing, it does not appear to capture the same information as the NASA-TLX or the intuition measures. In this study, the SUS only revealed significant differences between the CHC and the WYSWYE authentication methods; neither method's scores were significantly different than those of the PIN method. Both the CHC and the WYSWYE methods were unfamiliar to participants, and the objective data indicate that they were more difficult to use than the PIN method. Thus, the SUS results are surprising in some ways, as nearly half of the items should be sensitive to ease of use (e.g., ease of use, need for technical support, and system learnability). Even so, there are no items that directly measure familiarity, and most of the other items focus on the quality of the interaction experience (e.g., likelihood of future system use, confident use, interaction consistency, well-integrated functions). It may be the lack of items indexing familiarity that causes the SUS to be less sensitive to intuitiveness than the other measures.

Even though these are preliminary data, they do reinforce the idea that further consideration of the elements that define an intuitive experience are needed. What are the necessary components of intuition that must be captured in the subjective report? Further, how should those components of the interaction be weighted? For instance, should more emphasis be placed on familiarity? Automaticity? Subjective ease of use? If familiarity is to be measured, how can it be measured in a way that will capture implicit knowledge?

One of the recurrent themes in this chapter is the challenge of assessing the contributions of familiarity and implicit learning to the intuitive experience. From a design perspective, the phenomena of incidental and implicit learning have important implications. Users can learn about interactions in a variety of unexpected situations, expending little to no effort in the process. In addition, users can learn about regularities in the environment but have no conscious access to that information, even though they can apply the knowledge appropriately (e.g., Broadbent & Aston, 1978; Cleeremans & McClelland, 1990; Fried & Holyoak, 1984; Lewicki, Hill, & Bizot, 1988; Reber, 1967). In some regards, these types of learning are incredibly valuable because they support effective interactions. However, these same processes can be problematic when attempting to understand when and why an interaction is intuitive. Users may learn about interactions in a variety of situations that would be unexpected to the designer; this prevents designers from asking specific questions about those interaction experiences. Further, even if an individual is asked about a relevant interaction experience, they may not have a distinct memory of learning about the interaction, especially if they put no effort into that learning. It is possible that if the interaction was learned implicitly, the user may have no conscious access to the rules they are applying to resolve interaction ambiguity.

Similarly, the feeling of familiarity is vague, such that even if the user knows the interaction is familiar, he or she may not be able to verbalize why it seems familiar. Some measures of intuition ask users to indicate, essentially, the strength of their familiarity signal (Naumann & Hurtienne, 2010); other measures of intuition take a more indirect route, asking about relevant interactions that the user might have encountered in the past (e.g., Blackler et al., 2003). It is apparent that familiarity plays a critical role in intuitive interactions, but there is no agreed method for indexing that information. It may be that a global measure of the strength of the familiarity signal is the most accurate familiarity measure because the user does have conscious access to that information. But that measure does not necessarily provide actionable information for designing novel intuitive interactions. On the other hand, the approach taken by Blackler and colleagues may provide a better measure of the specific interactions or features that underlie that familiarity signal, but the challenge is finding a way to elicit the information while circumventing the memorial challenges (access to implicit knowledge and recollection of past technological experiences). For these reasons, much intuitive interaction research has involved coding the actual use of interfaces and prototypes. However, this is time-consuming and often not feasible during the product development cycle (Fischer et al., 2015)

3.7   CONCLUSION

In this chapter, we have identified cognitive processes that play a central role in supporting intuitive interactions. By exploring the characteristics of these processes, a more comprehensive understanding of what is required for an interaction to be intuitive can be realized. To help convey these cognitive principles and their relation to design, we created the Intuitive Interaction Hierarchy (Figure 3.1). In light of the various influences on intuition, we considered how knowledge representations might constrain the methods used to effectively elicit that knowledge (Still et al., 2015). For instance, interactions that are supported by consistent associations in memory are unlikely to be available for conscious report. Therefore, they may need to be assessed using indirect or implicit measures. We also considered what elements of intuitive interactions should be measured using subjective report.

In conclusion, it may be difficult in some cases to maintain an intuitive experience for users. But the cost of offering an intuitive experience may be well worth it. It is important to remember that intuitive experiences, usually supported by automatic processes and implicit learning, also allow limited cognitive resources to be reallocated to higher-level tasks. One of the primary goals of designers is to facilitate the user's ability to complete tasks effectively without having to be distracted by the supporting interface. Our hope is that a more nuanced understanding of cognitive processes and subsequent translation to the design community will help support the development of these intuitive interactions.

REFERENCES

Allen, B. G., & Buie, E. (2002). What's in a word? The semantics of usability. *Interactions*, 9, 17–21.
Andre, A. D., & Wickens, C. D. (1995). When users want what's not best for them. *Ergonomics in Design*, 3, 10–13.
Atkinson, R. C., & Juola, J. F. (1974). Search and decision processes in recognition memory. In D. H. Krantz, R. C. Atkinson, and P. Suppes (Eds.), *Contemporary Developments in Mathematical Psychology* (pp. 243–290). San Francisco, CA: Freeman.
Bailey, R. W. (1993). Performance vs. preference. In *Proceedings of the 37th Meeting of the Human Factors and Ergonomics Society* (pp. 282–286). Seattle, WA: HFES.

Bergum, B. O., & Bergum, J. E. (1981). Population stereotypes: An attempt to measure and define. *In Proceedings of the 25th Meeting of the Human Factors and Ergonomics Society* (pp. 662–665). HFES.

Besnard, D., & Cacitti, L. (2005). Interface changes causing accidents: An empirical study of negative transfer. *International Journal of Human–Computer Studies*, 62, 105–125.

Betsch, C. (2008). Chronic preferences for intuition and deliberation in decision making: Lessons learned about intuition from an individual difference approach. In H. Plessner, C. Betsch, and T. Betsch (Eds.), *Intuition in Judgment and Decision Making* (pp. 231–248). New York: Lawrence Erlbaum.

Betsch, T. (2008). The nature of intuition and its neglect in research on judgment and decision making. In H. Plessner, C. Betsch, and T. Betsch (Eds.), *Intuition in Judgment and Decision Making* (pp. 3–22). New York: Lawrence Erlbaum.

Blackler, A. (2008). *Intuitive Interactions with Complex Artifacts: Empirically-Based Research.* Saarbrucken, Germany: VDM.

Blackler, A., Desai, S., McEwan, M., Popovic, V., & Diefenbach, S. (2018). Perspectives on the nature of intuitive interaction. In A. Blackler (Ed.), *Intuitive Interaction: Research and Application* (pp. 19–40). Boca Raton, FL: CRC Press.

Blackler, A., Popovic, V., & Desai, S. (2018). Intuitive interaction research methods. In A. Blackler (Ed.), *Intuitive Interaction: Research and Application* (pp. 65–88). Boca Raton, FL: CRC Press.

Blackler, A., Popovic, V., & Mahar, D. (2003). The nature of intuitive use of products: An experimental approach. *Design Studies*, 24, 491–506.

Blackler, A., Popovic, V., & Mahar, D. (2010). Investigating users' intuitive interaction with complex artifacts. *Applied Ergonomics*, 41, 72–92.

Broadbent, D. E., & Aston, B. (1978). Human control of a simulated economic system. *Ergonomics*, 21, 1035–1043.

Brooke, J. (1996). SUS: A quick and dirty usability scale. In P. W. Jordan, B. Thomas, B. A. Weerdmeester, and A. L. McCelland. *Usability Evaluation in Industry.* London: Taylor and Francis.

Cleary, A. M., Langley, M. M., & Seiler, K. R. (2004). Recognition without picture identification: Geons as components of the pictorial memory trace. *Psychonomic Bulletin and Review*, 11, 903–908.

Cleary, A. M., Ryals, A. J., & Nomi, J. S. (2013). Intuitively detecting what is hidden within a visual mask: Familiar–novel discrimination and threat detection for unidentified stimuli. *Memory and Cognition*, 41, 989–999.

Cleeremans, A. & McClelland, J. L. (1991). Learning the structure of event sequences. *Journal of Experimental Psychology: General*, 120, 235–253.

Epstein, S. (2010). Demystifying intuition: What it is, what it does, and how it does it. *Journal of Psychological Inquiry*, 21, 295–312.

Ericsson, K. A., & Simon, H. A. (1993). *Protocol Analysis: Verbal Reports as Data.* Cambridge, MA: MIT Press.

Fischer, S., Itoh, M., & Inagaki, T. (2015). Screening prototype features in terms of intuitive use: Design considerations and proof of concept. *Interacting with Computers*, 27, 256–270.

Fried, L. S., & Holyoak, K. J. (1984). Induction of category distributions: A framework for classification learning. *Journal of Experimental Psychology: Learning, Memory, and Cognition*, 10, 234–257.

Gibson, J. J. (1979). *The Ecological Approach to Visual Perception.* Boston, MA: Houghton Mifflin.

Hart, S. G. (2006). NASA Task Load Index (NASA-TLX): 20 Years Later. In *Proceedings of the 50th Meeting of the Human Factors and Ergonomics Society* (pp. 904–908). Santa Monica, CA: HFES.

Hart, S. G., & Staveland, L. E. (1988). Development of NASA-TLX (Task Load Index): Results of empirical and theoretical research. In P. A. Hancock and N. Meshkati (Eds.), *Human Mental Workload*. Amsterdam, the Netherlands: North Holland Press.

Hodgkinson, G. P., Langan-Fox, J., & Sadler-Smith, E. (2008). Intuition: A fundamental bridging construct in the behavioral sciences. *British Journal of Psychology*, 99, 1–27.

Hurtienne, J., & Israel, J. H. (2007). Image schemas and their metaphorical extensions: Intuitive patterns for tangible interaction. In *Proceedings of the 1st International Conference on Tangible and Embedded Interaction* (TEI) (pp. 127–134), Baton Rouge, LA: ACM.

ISO/IEC. (2003). International Standard 18035: Information technology; Icon symbols and functions for controlling multimedia software applications (1st ed.).

Keppel, G., & Underwood, B. J. (1962). Proactive inhibition in short-term retention of single items. *Journal of Verbal Learning and Verbal Behavior*, 1, 153–161.

Khot, R. A., Kumaraguru, P., & Srinathan, K. (2012). WYSWYE: Shoulder surfing defense for recognition based graphical passwords. In *Proceedings of the 24th Australian Computer–Human Interaction Conference* (pp. 285–294). New York: ACM.

Lewicki, P., Hill, T., & Bizot, E. (1988). Acquisition of procedural knowledge about a pattern of stimuli that cannot be articulated. *Cognitive Psychology*, 20, 23–37.

Lewis, J. R., & Sauro, J. (2009). The factor structure of the system usability scale. In *Proceedings of the 1st International Conference on Human Centered Design* (pp. 94–103). Berlin, Germany: Springer.

Logan, G. D. (2002). An instance theory of attention and memory. *Psychological Review*, 109, 376–400.

Matthies, D. J. C., Muller, F., Anthes, C., & Kranzlmuller, D. (2013). ShoeSoleSense: Proof of concept for a wearable foot interface for virtual and real environments. In *Proceedings of the 19th ACM Symposium on Virtual Reality Software and Technology* (pp. 1–4), Singapore: ACM.

Mandler, G. (1980). Recognizing: The judgment of previous occurrence. *Psychological Review*, 87, 252–271.

Manza, L., & Bornstein, R. F. (1995). Affective discrimination and the implicit learning process. *Consciousness and Cognition*, 4, 399–409.

Naumann, A., & Hurtienne, J. (2010). Benchmarks for intuitive interaction with mobile devices. In *Proceedings of the 12th International Conference on Human Computer Interactions with Mobile Devices and Services* (pp. 401–402). Lisbon, Portugal.

Naumann, A., Hurtienne, J., Israel, J. H., Mohs, C., Kindsmüller, M. C., Meyer, H. A., & Hublein, S. (2007). Intuitive use of user interfaces: Defining a vague concept. In D. Harris (Ed.), *Engineering Psychology and Cognitive Ergonomics* (pp. 128–136). Heidelberg, Germany: Springer.

Norman, D. A. (1988). *The Psychology of Everyday Things*. New York: Basic Books.

Norman, D. A. (1999). Affordance, conventions, and design. *Interactions*, 6, 38–42.

Norman, D. A., & Shallice, T. (1986). Attention to action: Willed and automatic control of behavior. In R. J. Davidson, G. E. Schwartz, & D. Shapiro (Eds.), *Consciousness and Self-Regulation: Advances in Research and Theory*, Vol. 4 (pp. 1–18). New York: Plenum Press.

Posner, M. I., & Snyder, C. R. R. (1975). Facilitation and inhibition in the processing of signals. In P. M. A. Rabbitt & S. Dornic (Eds.), *Attention and Performance V* (pp. 669–682). New York: Academic Press.

Raskin, J. (1994). Intuitive equals familiar. *Communications of the ACM*, 37, 17–18.

Reber, A. S. (1967). Implicit learning of artificial grammars. *Journal of Verbal Learning and Verbal Behavior*, 6, 855–863.

Reber, A. S. (1989). Implicit learning and tacit knowledge. *Journal of Experimental Psychology: General*, 118, 219–235.

Rumelhart, D. E., & Norman, D. A. (1982). Simulating a skilled typist: A study of skilled cognitive-motor performance. *Cognitive Science*, 6, 1–36.

Shiffrin, R. N., & Schneider, W. (1977). Controlled and automatic human information processing, II: Perceptual learning, automatic, attending, and a general theory. *Psychological Review*, 84, 127–190.

Simon, H. A. (1969). *The Sciences of the Artificial* (2nd ed). Cambridge, MA: MIT Press.

Spool, J. M. (2005). What makes a design seem "intuitive"? User Interface Engineering. Online: http://wwwuie.com/articles/design_intuitive (published January 10).

Still, J. D., & Dark, V. J. (2008). An empirical investigation of affordances and conventions. In J. S. Gero and A. K. Goel (Eds.), *Design Computing and Cognition* (pp. 457–472). Dordrecht, the Netherlands: Springer.

Still, J. D., & Dark, V. J. (2013). Cognitively describing and designing affordances. *Journal of Design Studies*, 34, 285–301.

Still, J. D., Still, M. L., & Grgic, J. (2015). Designing intuitive interactions: Exploring performance and reflection measures. *Interacting with Computers*, 27, 271–286.

Topolinski, S., & Strack, F. (2009). The architecture of intuition: Fluency and affect determine intuitive judgments of semantic and visual coherence and judgments of grammaticality in artificial grammar learning. *Journal of Experimental Psychology: General*, 138, 39–63.

Wiedenbeck, S., Waters, J., Sobrado, L., & Birget, J-C (2006). Design and evaluation of a shoulder-surfing resistant graphical password scheme. In *Proceedings of AVI 2006* (pp. 177–184). Venice, Italy: ACM.

Wippish, W. (1994). Intuition in the context of implicit memory. *Psychological Research*, 56, 104–109.

Ullrich, D., & Diefenbach, S. (2010a). INTUI: Exploring the facets of intuitive interaction. In J. Ziegler and A. Schmidt (Eds.), *Mensch und Computer* (pp. 251–260). Munich, Germany: Oldenbourg.

Ullrich, D., & Diefenbach, S. (2010b). From magical experience to effortlessness: An exploration of the components of intuitive interaction. In *Proceedings of the 6th Nordic Conference on Human–Computer Interaction: Extending Boundaries* (pp. 801–804). New York: ACM.

You, H., & Chen, K. (2007). Applications of affordances and semantics in product design. *Design Studies*, 28, 23–28.

Part II

---

*Research and Findings in
Intuitive Interaction*

4 Research Methods for Intuitive Interaction

Alethea Blackler, Vesna Popovic, and Shital Desai

CONTENTS

4.1 INTRODUCTION

As intuitive interaction research has grown over the past 18 years, researchers have had to develop the best ways to go about understanding this phenomenon. This chapter explains the ways researchers have been able to define the concept of intuitive interaction by carefully designing their research and selecting appropriate data collection methods, and explores which research methods are now being employed in intuitive interaction research around the world. Finally, it presents a toolkit of appropriate methods that can be used by researchers to further their study on intuitive interaction. The Intuitive Interaction Research Methods Toolkit summarizes the approaches followed by researchers all over the world that have been successful in achieving robust findings, as detailed in this chapter (Table 4.1). The toolkit is categorized into four areas: (1) data collection methods, (2) objectives, (3) goals, and (4) measures. The toolkit offers a way for designers and researchers to identify the methods that they can use in their investigation once the objectives and goals of their study have been confirmed.

Much research suggests that intuition relies on experiential knowledge (Agor, 1986; Bastick, 2003; Bowers, Regehr, Balthazard, & Parker, 1990; Dreyfus, Dreyfus, & Athanasiou, 1986; Fischbein, 1987; King & Clark, 2002; Klein, 1998; Noddings & Shore, 1984). In addition, researchers agree that intuition is a process that is not consciously available as it takes place, so details of intuitive processing are not subjectively known (Agor, 1986; Bastick, 2003). The nature of intuition itself, being unconscious, makes it challenging to study. Few experiments have been conducted that specifically target intuition (Bastick, 2003), so there was no established method specifically aimed at studying intuitive interaction when research in this field began at the turn of the twenty-first century.

The Intuitive Use of User Interfaces (IUUI; Germany) definition of intuitive use was developed based on a literature review of usability design criteria and a series of interviews and workshops with users, usability specialists, and user interface design practitioners. (For an overview of the results of these workshops, see Mohs et al., 2006). We, the Queensland University of Technology (QUT) group in Australia, used a literature review on intuition and usability, interaction design, human-centered design, and human factors to reach our definition (Blackler, Popovic, & Mahar, 2010; Blackler, Popovic, & Mahar, 2003). Both the IUUI and QUT defined intuitive interaction as a process based on past experience, applied unconsciously, and contributing to the fast and successful use of interfaces. The definitions were empirically tested by both groups using experiments designed to establish whether they were supported.

Much intuitive interaction research has taken an empirical experiment-based approach. Perhaps due to the general misunderstanding of the term *intuitive* and its overuse in marketing literature, researchers have felt the need to show empirically what intuitive interaction really is and how it can be facilitated. Generally, these experiments have focused on generating and reporting quantitative results even though some of the raw data collected have been qualitative (e.g., videos of people performing tasks). Many researchers in this field have been careful to make sure they have appropriate experiment designs and sample sizes for statistical analysis as well as rigorous measures of intuitive interaction. From our point of view, the main reason for this is that the most successful way to investigate such rich and complex data was believed to be through statistical analysis. Also, due to the novel nature of this research, we felt that quantitative measures would be required to support the claims made.

Experiments have fallen into four categories. First are those intended to understand intuitive interaction itself and confirm whether the definitions previously developed were accurate; generally, this covers the early work by the QUT and IUUI groups (e.g., Blackler et al., 2010; Blackler et al., 2003; Hurtienne & Blessing, 2007). Second were those intended to test interfaces against each other to see whether some were more intuitive than others, using similar methods to those developed initially to compare interfaces for intuitive interaction (Gudur, Blackler, Popovic, & Mahar, 2009; Macaranas, Antle, & Riecke, 2015; Mihajlov, Law, & Springett, 2014; Reddy, Blackler, Popovic, & Mahar, 2011) and interaction modalities (Chattopadhyay & Bolchini, 2015). Third were those intended to test interfaces developed using tools for designing for intuitive interaction. A number of experiments have been conducted to test interfaces developed to be more intuitive using such tools. These tests have enabled us to understand whether designing for intuitive interaction according to the theory and recommendations developed can be effective or not. For example, Hurtienne, Klöckner, Diefenbach, Nass, and Maier (2015) reported on a study where they applied image schemas to an interface during the design process and tested the resulting interface with older people, and we (Blackler, Popovic, & Mahar, 2014) tested an interface redesigned using our tool for intuitive interaction against the original (commercial) interface. Fourth are those experiments intended to see what effects age has on intuitive interaction and to investigate the differences in intuitive interaction between age groups (e.g., Blackler, Popovic, Mahar, Reddy, & Lawry, 2012; O'Brien, 2018), as described in Chapter 1 (Blackler, 2018).

4.2 OBSERVATION APPROACH

The methods used by QUT and others to conduct the experiments in all four of these categories have been various, but the underpinning approach has been the quantification of people performing tasks with real or prototyped interfaces. There are many aspects of interaction that cannot be accessed through asking opinions and feedback; we need to watch real people doing real things to really understand what they do. Several other methods (e.g., surveys, diaries, questionnaires, interviews, and verbal protocols) have been successfully used to supplement the actual participant activities and will be discussed in subsequent sections, but these have not been relied on solely in most cases. For example, in the initial experiments conducted by us (Blackler et al., 2010), participants were video-recorded performing set tasks with products and interfaces while performing concurrent (think-aloud) protocol. Then they were interviewed about their familiarity with each interface feature they used.

4.2.1 MEDIATING INTERFACES AND PRODUCTS

Physical artifacts are mediators between the agent and the activity of manipulating these artifacts (Nardi, 1996). This manipulation of artifacts brings out natural behavior in the agent (Engeström, Miettinen, & Punamäki, 1999). Thus, artifacts can be used to study people's complex behaviors (Popovic, 2003), as they result in human activities that are representative of user goals. Many of the experiments described here have used real products, including cars (Lewis, Langdon, & Clarkson, 2008), e-readers, video cameras (O'Brien, 2010), alarm clocks (Lawry, Popovic, & Blackler, 2011; O'Brien, 2010), remote controls, cameras (Blackler et al., 2010; Lawry et al., 2011; Lewis et al., 2008), toys (Desai, Blackler, & Popovic, 2015, 2016), websites (Mohan, Blackler, & Popovic, 2015), video games (McEwan, Blackler, Johnson, & Wyeth, 2014), microwaves (Lewis et al., 2008), airports (Cave, Blackler, Popovic, & Kraal, 2014), and various other domestic products chosen by participants (Lawry, Popovic, & Blackler, 2009, 2010). However, prototyped or mocked-up interfaces have also been used, such as microwaves (Blackler et al., 2014), accounting software

(Hurtienne, 2009), music systems (Hurtienne et al., 2015), onboard vehicle systems (Fischer, 2018; Fischer, Itoh, & Inagaki, 2015), games of various sorts on touch-screens or with full-body interaction (Gudur, Blackler, Popovic, & Mahar, 2013; Macaranas et al., 2015; Mihajlov et al., 2014), health-care devices (Gudur et al., 2009), and finally interfaces designed specifically to test ideas about intuitive inter-action that were not intended to be full or partial prototypes of any product or system but were designed to allow participants to go through procedures that would reveal answers about their interaction (Chattopadhyay & Bolchini, 2015; Still & Still, 2018; Tretter, Diefenbach, & Ullrich, 2018).

In many experimental situations, prototyping is necessary as it allows research-ers to test differences between interface designs, but it is not without its problems. Touchscreen prototypes have been employed in several of our experiments. Issues encountered include difficulties in creating suitable software prototypes, represent-ing and conveying a three-dimensional object in two dimensions, prototype and interface sizing and scaling, and representations of LCD displays on a touchscreen (Blackler, 2008b). Strategies to get around the kinds of issues encountered can work successfully (e.g., see Blackler, 2008b) but need to be well thought out and well tested. Hawthorn (2007) also notes that older people find it difficult to work with low-fidelity prototypes and suggests making testing environments as realistic as pos-sible. Hence, we have used only high-fidelity prototypes.

4.2.2 TASKS

When asking participants to use such an interface, researchers give them tasks to complete during the session. Tasks can be seen as substitutes for real user goals, and it is important that selected tasks are realistic, as unrealistic tasks may lead to unre-alistic behavior (Vermeeren, 1999). They must involve the use of all features that are under investigation, must accurately simulate real product use, and must be relevant to the user (McClelland, 1995). Most importantly, researchers can choose which fea-tures participants use in the interface by designing the tasks. Therefore, participants can be required to use the parts of the product most likely to yield results about the intuitive use of the features researchers are interested in.

4.3 RELEVANT MEASURES FOR INTUITIVE INTERACTION

All of the experiments conducted in intuitive interaction have faced similar chal-lenges in identifying and quantifying when participants were using interface features intuitively, as well as what interface features, activities, and ideas participants they were recruiting may have already been familiar with, which obviously would dictate what past experience they were bringing to the interaction. Finding a way to mea-sure variables as nebulous as technology familiarity (TF) and intuitive interaction has been a challenge. This has been met in various ways over the years so that there now exists a well-tested toolkit of options for understanding and testing intuitive interactions, both objective (e.g., researcher coded or automatically collected) and subjective (generally based on user/participant ratings or feedback), and quantifying for TF or relevant past experience (generally subjectively rated). The next sections will discuss how these challenges have been addressed by various researchers, and the measurement tools that have been developed.

4.3.1 Measuring Familiarity

Researchers agree that familiarity is essential for intuitive interaction, so a challenge facing researchers in this area has been identifying the familiarity of participants. TF or a very similar variable has been used multiple times (Blackler et al., 2011; Cave et al., 2014; Desai et al., 2015; McEwan et al., 2014; O'Brien, 2010; Still & Still, 2018). It has been measured through a questionnaire in which participants provided details of their experience with products with similar features to those they would encounter during the experiment. A separate questionnaire has therefore been devised for each new product used in our experiments according to the features we were interested in exploring. We asked them how often they used each relevant products, from "never" to "every day," and how many of the features they used, from "all of them" (including reading the manual) to "none." More frequent and more extensive use of the products and features produces a higher TF score (Blackler et al., 2010).

Hurtienne, Horn, and Langdon (2010) discussed the facets of prior experience and identified exposure, competence, and subjective feeling. Exposure is split into three subcomponents: duration, intensity, and diversity of use. Duration describes how many months/years a product has been used, intensity describes how often it is used, and diversity of use describes the number of functions/features used or problems solved with the product. In addition, Hurtienne et al. (2010) suggest that three levels of specificity apply to exposure and competence. These are (1) the product in focus, (2) different products of the same type, and (3) a broad range of products of different types. Our TF questionnaires were designed to measure exposure (intensity and diversity) at all three levels of specificity, as interface features are often transferred between products and product types. For example, the universal remote control we used for some of our early experiments (Blackler et al., 2010) had tab-based navigation similar to software alongside standard audiovisual symbols found on all kinds of remote controls and audiovisual devices.

McEwan et al. (2014) developed a more precise version of our TF questionnaire, Game Technology Familiarity (GTF), for research into intuitive interaction with video games. We have found that adding images to the questionnaire (as per Desai et al., 2015; Gudur et al., 2009; McEwan et al., 2014) has helped to make it more self-explanatory so that it can be delivered online without support and with more confidence than previously. O'Brien also used a version of our TF questionnaire (O'Brien, 2018). She was interested in assessing general technology experience, in order to identify high-tech and low-tech older people for her experiments and diary studies. This experience not only indicates familiar technologies but also specific features and controls, interaction techniques, expected guidance and feedback, and typical problem-solving techniques for these technologies.

Lawry et al.'s two studies (Lawry et al., 2009, 2010, 2011) examined familiarity and prior knowledge, and methods to elicit relevant prior knowledge from participants that would be suitable for use in the design industry context. Some new ideas were developed using interviews and a measure known as *task recall*. Task recall involves participants using products as prompts but not operating them, instead talking through how they would complete common tasks in order to assess how well they knew them. Task recalls were recorded and coded for familiarity based on similar heuristics to those used by us previously, focusing particularly on accuracy, and also on whether participants were chunking sets of subtasks together (which could imply high familiarity).

4.3.2   OBJECTIVE MEASURES

Objective data collected from these types of experiments have included outputs from coding video of participants completing tasks for accurate uses, errors, and intuitive uses, or counts commonly used in usability research such as the number of clicks to complete tasks, latency, or time to complete tasks. Automated software has sometimes been used to collect some performance data such as accurate uses and reaction times.

Accuracy or error count is a relevant measure, as intuition is generally acknowledged to be fairly accurate in providing solutions. Most intuition researchers agree that intuition is a useful guide that rarely misleads (Bastick, 2003). Time on task is also relevant to quantifying intuitive interaction, as intuitive processing is assumed to be faster than more conscious types of processing (Agor, 1986; Bastick, 2003; Salk, 1983), so participants interacting intuitively with a product should be able to complete tasks more quickly. Thus, many researchers in intuitive interaction have used time to complete tasks or subtasks, latency and correctness of uses, error counts, or successful task completion in order to help identify intuitive interaction (Brandenburg & Sachse, 2012; McEwan et al., 2014; Reddy et al., 2011; Reddy, Blackler, Popovic, & Mahar, 2009). Other objective measures conceptually linked to the speed and success of intuition and often used in usability testing and similar research have been used in intuitive interaction research, such as percentage complete (Desai et al., 2015; McEwan et al., 2014), the NASA Raw Task Load Index (NASA-RTLX) cognitive workload questionnaire (O'Brien, 2018), and the System Usability Scale (SUS) (Still & Still, 2018). Other researchers have also applied established measures and data collection procedures from psychology, such as dual-task protocols, in order to understand more about how intuition powers intuitive interaction (e.g., Fischer et al., 2015; Still, Still, & Grgic, 2015) and to find ways to apply it to and test interfaces more easily and efficiently. For example, Fischer (2018) developed her *screening method* and *false belief technique* based on psychology research protocols (dual-task and Deese–Roediger–McDermott [DRM], respectively). Reinhardt and Hurtienne (2017) are investigating the utility of a method which utilizes a dual-task (foot tapping along set tasks during usability testing) in an attempt to reduce the analysis required to identify intuitive interactions. They hope this method will reveal which usability issues are related to increased mental workload and hence are problems of intuitive interaction as opposed to problems of usability issues (such as physical fit).

Hurtienne used custom software to test the metaphorical extensions of image schemas applied to interface design. The software automatically measured the dependent variables: accurate responses to dialogue box instructions and response times. Hurtienne and Blessing (2007) conducted two empirical experiments to investigate the application of image schema theory to user interface design to build better products in terms of effectiveness, (mental) efficiency, and user satisfaction. Image schemas are "abstract representations of recurring dynamic patterns of bodily interactions that structure the way we understand the world" (Hurtienne & Blessing, 2007, p. 130) and are thus important building blocks for thinking. They are based on each individual's experience of interaction with the physical world but tend to be largely universal, as the physical world operates in the same way for almost everyone who is able-bodied. Because they are based on past experience and because they are so well known and so universal that they become unconscious, image schemas can be defined as intuitive. Therefore, Hurtienne argued, incorporating image schemas into interfaces can allow intuitive interaction. The experiments were conducted in three steps: (1) suitable metaphorical extensions of image schemas were identified, (2) user interface elements were laid out consistently or inconsistently with the metaphor, and (3) experiments were executed with users to measure the usability variables. The metaphorical extensions MORE IS UP, GOOD IS UP, and VIRTUE IS UP were investigated with 40 participants in experiment 1. Participants were asked to enter data from a fictitious survey of 20 hotels into a simulated interface. The survey results were presented to participants in the form of sentences such as "Staff is friendly." The sentences were aimed at priming the participants toward the metaphorical extensions of the image schemas. The participants were then asked to select buttons on the interface. The buttons were arranged at the UP and DOWN positions on the interface to be either compatible or incompatible with the metaphors MORE, GOOD, and VIRTUE. The hypothesis was that the participants should take less time to respond to the compatible arrangement than to the incompatible arrangement. The incompatible arrangement should also lead to more errors on the part of the participants. The authors conducted a second experiment to investigate the same hypothesis as in the first experiment but focused on two types of metaphorical extensions: quantitative (MORE IS UP) and qualitative (GOOD IS UP).

Fischer et al. (2015) conducted an experiment consisting of two phases: study and usage. A stimulus was presented to the users in the study phase, which was followed by a usage phase in which the participants are expected to perform actions on the interface. Multistate interfaces are interfaces that have numerous features and several of these features are available at one time. Configurations of several features in an interface that are available to users to interact with are referred to as *states*. The outcomes of both the phases were entered into log files for analysis. In the study phase, screenshots were taken of all the states (i.e., several features were captured in one screenshot—e.g., each navigation page of a website). Word and function clues that matched each of the screenshots were developed, along with clues that did not match the screenshots. Word clues represented words that appeared or did not appear in the screenshot (i.e., matched or did not match), while function clues represented functions of the features in the states. A clue was presented to the participants for a predefined maximum time, and participants were given an option to continue within the maximum time by clicking on the GO button. The screenshot to be matched was then displayed and the participants were presented with YES and NO buttons to present their judgment on whether the clues were a match or a mismatch to the screenshots. In the usage phase, a set of tasks to be performed on the interface were prepared and presented to the participants. These tasks were goal specific, which required participants to navigate from the current state to another to achieve a certain goal. Participants were allowed to explore as many states or as many features in the interface as possible. Accuracy was considered to be inversely proportional to the number of states explored. The results of applying this schema induction paradigm revealed that the induction of a new schema improved exploration of the states for new features but not for familiar features. Familiar features were processed through the transfer of prior schemas, whereas those less familiar required induction for new schemas. The knowledge required to use the features intuitively is derived not only from across devices but also from across domains. Fischer (2018) has stated that products that tap into novel schema spaces tend to reveal errors as they force first-time users into task solving and deliberation. She states that performance with these types of products is therefore best reflected by task accuracy, whereas products that tap into a better-known space allow the transfer of existing schemas, meaning their first-time use is more likely to be intuitive, and performance is best measured through metrics such as decision times (e.g., latency), completion times, and automaticity (e.g., dual-task performance, as per Fischer et al., 2015).

Antle, Corness, and Droumeva (2009) used the following objective measures to determine the effectiveness of embodied metaphors in intuitive interaction: (1) time taken to complete a task, (2) accuracy of task completion, (3) accuracy of the verbal explanation for each subtask. Two researchers took notes about the discrepancies between the actions of the participants and their verbalization rating the verbal statements of their participants for ease of learning, intuitiveness of learning, and amount of concentration required to learn to use the system.

4.3.2.1   Coding for Intuitive Interaction

However, at least initially, on top of these relatively easy-to-quantify measures, there also needed to be a specific measure of intuition or intuitive uses. The challenge was to find ways of coding the video of observations so that this level of detail could be extracted from user data. Coding observational or verbal data requires specialist software if it is to be done reliably and within a reasonable timeframe. The Noldus Observer XT is a system for collecting, managing, analyzing, and presenting observational and audiovisual data. It captures a level of detail not possible in live situations and that cannot be analyzed easily without dedicated software. The program, like much data analysis software, requires a coding scheme to be set up. For example, the coding scheme employed by Blackler et al. (2010) assumed that various levels of cognitive processing occur during one task (Berry & Broadbent, 1988) and was designed to distinguish intuitive processing from other processes (e.g., automatic and conscious processes). A set of heuristics for identifying intuitive uses was developed and used to code the audiovisual data of participants completing tasks (Blackler et al., 2011).

The first heuristic was whether each use was conducted within a short timeframe, as intuition is generally fast (Agor, 1986; Bastick, 2003; Hammond, 1993; Salk, 1983) and the time taken to make a move (latency) can be used to measure thinking time (Cockayne, Wright, & Fields, 1999). If a participant had already spent some time exploring other features before hitting the correct one, that use was unlikely to be intuitive, so those uses coded as intuitive involved the participants using the right feature with no more than 5 s hesitation, commonly closer to 1 or 2 s (Blackler et al., 2010). The second heuristic was whether each feature use was approached with a reasonable certainty of correctness and a clear expectation about how a function would respond, because intuition is correct in that it harmonizes all the subjective information currently available and intuitive perceptions are experienced as true in the same way that sensory data are experienced as true (Bastick, 2003). The third heuristic was whether there was any mention by participants in the verbal protocol of having used a feature before. The fourth was whether there was evidence of non-conscious processing or conscious reasoning (Blackler et al., 2010). For example, the following is an excerpt of a transcript from a reasoning-based use.

I'll just experiment ... I'm not sure. This changes the screen so I'll change ... this is an arrow up so I'll change ... ahh ... demonstration ... ah ... language ... clock set. I've reached the dot by clock set so that's the point of that dot there. OK, so it looks as though I'm getting there.

And the following is an intuitive use of the same feature.

Aha! OK here we go and I want to go to clock set. OK.

These examples show that, although both participants were completing the same action, the level of reasoning is different for each. The second participant (the intuitive use) did not verbalize the steps required to get the marker to the clock set position, so the intuitive use lacks the detail of the reasoning process and shows that the participant understood how the system would respond to his actions. The first participant verbalized a systematic reasoning process and showed much less certainty about what he was doing and how the system would respond.

Intuitive use codes were applied cautiously, only when the use showed two or more of these characteristics and the researcher was certain about the type of use. At the early stages, every feature use was coded because it was the familiarity with and use of individual product features that was important, rather than the use of the product as a whole. In later experiments, these same heuristics have also been used to code larger chunks of video, comprised of small subtasks rather than each feature use (Gudur et al., 2009; O'Brien, 2010; Reddy et al., 2011). The advantage of this is that it saves time and allows the research to focus on aspects other than feature uses (e.g., participants' understanding of task procedures).

O'Brien (2010) used an observation study very similar to our early experiments. She video-recorded younger adults, high-TF older adults, and low-TF older adults completing tasks with three different contemporary products (a video camera, an e-reader, and a digital radio alarm clock). This was followed by an interview about the participants' familiarity with the features of the products and similar features they may have encountered elsewhere. She performed two levels of coding: one that coded every feature use as we did and one that looked at the performance of overall tasks. Her dependent variables were the correctness of feature uses (i.e., valid or invalid), references to information outside of their own experience (e.g., help from product labels or from the researcher), and task performance (i.e., optimal, successful, or partial).

4.3.3 SUBJECTIVE MEASURES OF INTUITIVE INTERACTION

Hurtienne and Naumann and Hurtienne (2010) and Naumann et al. (2007) investigated the subjective consequences of intuitive interaction. They suggested the effectiveness of interactions, the mental efficiency of the users, and user satisfaction as measures of intuitive interaction. They developed a questionnaire, the Questionnaire for the Subjective Consequences of Intuitive Use (QUESI), that covers all the measures of subjective consequences. The questionnaire consists of 14 items, grouped into five subscales: subjective mental workload, perceived achievement of goals, perceived effort of learning, familiarity, and perceived error rate. The score of each subscale is computed by evaluating the means of all the responses to the items of that subscale. The total score of the questionnaire is equal to the mean evaluated across all five subscales. Naumann and Hurtienne (2010) and Naumann et al. (2007) used QUESI to test mobile devices and applications for intuitive interaction. They found that devices and applications that users were familiar with scored higher QUESI scores than those that users were not familiar with.

Ullrich and Diefenbach (2010a) approached intuitive interaction as a process originating from "gut feeling" rather than from "reason." Accordingly, they suggested subjective and experiential components of intuitive interaction: *gut feeling* and *magical experience*. While gut feeling refers to the process of decision making, magical experience refers to the results of intuitive decision making. In addition to these two components, they agreed with two parameters of intuitive interaction suggested in the literature: effortlessness and verbalizability (Blackler, 2008a; Naumann & Hurtienne, 2010). Ullrich and Diefenbach (2010b) developed the INTUI questionnaire to assess these four components of intuitive interaction using a set of 16 questions. The questions were designed using the semantic differential technique, where participants were asked to answer each question on a seven-point scale between two bipolar statements: for example, "Using the product was inspiring" versus "Using the product was insignificant." The factorial structure of the INTUI questionnaire was tested in numerous studies (Ullrich & Diefenbach, 2010a, 2010b), and it has been employed in a large survey-based study using only subjective feedback (Diefenbach & Ullrich, 2015; Tretter et al., 2018). The INTUI questionnaire also included an additional measure to evaluate the overall perceived intuitiveness of the interaction using a seven-point scale.

McEwan et al. (2014) used the Player Experience of Need Satisfaction (PENS) and Game Experience Questionnaire (GEQ) measures, commonly used in games research, to collect subjective feedback from participants, especially the three-item Intuitive Controls component that forms a part of the PENS. McAran (2018), in his research incorporating perceived intuitiveness in the Technology Acceptance Model (TAM), has used common TAM measures. Self-reported use was employed with a selection of six commonly used legal technology products as the basis of responses. The degree of use, degree of feature use, and degree of voluntary use were collected using slide bar scales, whereas Likert scales were used to measure the perceived ease of use, perceived usefulness and compatibility (also standard TAM measures), and a unique perceived intuitiveness measure.

Fischer (2018) gave participants questionnaires to gauge the potential subjective impact of the interaction they experienced during her experiments, including AttrakDiff, an instrument for conducting longitudinal evaluations of user experience (Hassenzahl, Burmester, & Koller, 2008) and the Net Promoter Score (NPS; Satmetrix Systems, Bain & Company, & Reichheld, 2017), which is popular in industry for tracking product loyalty and was also used by Palmer, Ogunyoka, and Hammond (2018).

Antle et al. (2009) used the following subjective measures: perceived competence and individual statements related to the ease of learning, the intuitiveness of learning, and the amount of concentration required to learn. Participants were asked to complete a questionnaire at the end of each session. The questionnaire was based on the Intrinsic Motivation Inventory subscales "Enjoyment and Interest" and "Perceived Competence" (Ryan, 2006; Xie, Antle, & Motamedi, 2008). Participants were finally asked what they liked and disliked about their experience with the system, the response to which was recorded and noted. Macaranas, Antle, and Riecke (2015) used a seven-level score of perceived intuitiveness in the evaluation of intuitive interaction with a whole-body system.

4.3.4 OBJECTIVE VERSUS SUBJECTIVE MEASURES

Still and Still (2018) conducted a pilot experiment to examine the effectiveness of traditional usability measures at indexing intuitive interactions in comparison with instruments designed specifically to measure intuitive interactions. This included the NASA Task Load Index (NASA-TLX; Hart & Staveland, 1988) and the SUS (Brooke, 1996), the QUESI (Hurtienne & Naumann, 2010), and a questionnaire inspired by Blackler et al.'s (2010) TF questionnaire. They found that the QUESI was the most sensitive measure and the SUS the least. They suggest that, due to the different familiarity and workload/ease of use aspects measured by each instrument, the NASA-TLX in conjunction with some reliable index of familiarity could be used to measure intuitive interaction. It is interesting to note that the SUS, commonly used in usability testing, does not appear to be an effective measure of intuitive interaction. Further work needs to be conducted to confirm these findings and to determine the most appropriate suite of measures for intuitive interaction research in various contexts.

Another aspect to consider when collecting subjective data is what knowledge participants may have access to. Knowledge acquired implicitly is generally not consciously available and cognitive signals of familiarity may be vague (Still & Still, 2018), so people do not always "know what they know." For example, Still and Dark (2008) used both the observation of set tasks and subjective feedback to determine the mapping of buttons to actions. They showed through their experiment that if they had only used subjective feedback they would have mapped some of the buttons in the opposite direction of the existing convention revealed by actual use. Palmer et al. (2018) also found that survey results were more positive than those revealed through verbal protocol during actual use, and if only surveys were used, the results would suggest a different story about the target populations' satisfaction. This may make subjective data less reliable, although there are measures that can be taken to mitigate these effects; for example, we have more recently added images to TF questionnaires to help prompt people about features they do not remember they know or which are difficult to describe in words that laypeople understand (Gudur et al., 2009; McEwan et al., 2014). The alternative is to run through the TF questionnaire with each participant face to face to check everything was clear and there is nothing they have forgotten. This has been done by us (Blackler et al., 2010) and by O'Brien (2010).

In addition, when rating the intuitiveness of an interaction after the fact, impressions at the beginning and the end of an interaction are over-proportionally weighted (Tretter et al., 2018). This primacy/recency effect is a typical example for cognitive biases in human perception (Mayo & Crockett, 1964). The retrospective nature of such measures, combined with accompanying unconscious processes, can make them prone to inaccuracy (Tretter et al., 2018).

Diefenbach and Ullrich (2015), Tretter et al. (2018), and McAran (2018) are the only studies in this field that have only collected data via an online survey where participants reflected on their use of existing products, and so depended entirely on subjective feedback rather than any kind of task completion or observation. Most authors in the field have used mixed methods involving objective and subjective measures, and some have expressed reservations about relying solely on subjective data. For example, Palmer et al. (2018) found discrepancies in participant ratings based on culture, and others have also shown that subjective feedback about a product can be biased by cultural behaviors and beliefs (Bignami-Van Assche, Reniers, & Weinreb, 2003; Dell, Vaidyanathan, Medhi, Cutrell, & Thies, 2012). Still and Still (2018) also discovered that their objective measures based on actual behavior showed quite different results from subjective ratings alone, as did Palmer et al. (2018). Therefore, it is likely that, when dealing with an unconscious process such as intuition, much research will continue to be based on measurable behavior, supplemented with subjective feedback and opinions.

4.4   OTHER DATA COLLECTION METHODS

This section discusses the application of various other data collection methods in intuitive interaction research. None of these are unique to this research community, but they have been successfully adapted to serve the needs of researchers in intuitive interaction. Most of the time these have been used in conjunction with the observation and quantification of task performance.

4.4.1   VERBAL PROTOCOLS

The protocol method is widely accepted in the research community. It yields data that are unstructured and very rich, and flexible analytical methods can be used. Concurrent protocols (also known as *think-aloud* and *talk-aloud* verbal protocols) have been successfully used in empirical studies of intuitive interaction as they offer an immediate account of thoughts and actions either without or with minimal prompts (e.g., Blackler et al., 2010; Palmer et al., 2018). However, we have also used retrospective protocols on occasion (Desai et al., 2015; Lawry et al., 2011), with participants reviewing footage of themselves using interfaces and talking through what they did and why. Co-discovery (or iterative protocol) involves having two users undertake a task and encouraging them to talk to each other as they do so. This method is often not appropriate when exploring an individual's past experience of a feature, as interaction with another person could prevent a participant from revealing their own relevant past experience of the product features and using their own intuition. However, we did find it useful when working with children in retrospective interviews and in observations (Desai et al., 2015) as they were challenged by concurrent protocols, and having two children playing games together was more naturalistic.

Concurrent protocols can be difficult for children when they are interacting with systems that require a heavy cognitive load (Höysniemi, Hämäläinen, & Turkki, 2003). Children's cognitive abilities are not fully developed until their late teens, and this influences their ability to think out loud. For example, they have a limited ability to think simultaneously about more than one concept, such as playing with a toy and thinking aloud at the same time. Concurrent protocols make the conditions in an experiment unnatural and are difficult to facilitate in a field setting, such as a school, or in a collaborative context, such as gameplay (Hertzum, 2016). Trial experiments with children revealed that although instructed to think aloud, they remained engrossed in play. They were not clear about what to do when they were instructed to think or talk aloud. When prompted to verbalize their feelings and thoughts, they stopped playing and talked. They also talked about things that were irrelevant to the game. When they went back to play, their game strategies were different to those they employed before the interruption. When delivering protocol, they took longer to finish the game, reacting slowly to various stimuli. They also spent a lot more time in distributed visual behavior, looking for something to talk about. This confirmed that concurrent protocols affect children's performances of spatial tasks (Gilhooly, Fioratou, & Henretty, 2010), which makes them difficult to use in the context of play.

Considering the problems with concurrent protocols, iterative protocol or co-discovery (Lim, Ward, & Benbasat, 1997) was used. Each child participant was paired with another to solicit in-depth information and free-flowing discussion between them (Adebesin, De Villiers, & Ssemugabi, 2009). The verbalization of ongoing thought processes provided direct insights into the knowledge and methods used during the play (Popovic, Kraal, Blackler, & Chamorro-Koc, 2012). Co-discovery encouraged children to question each other and to engage in deeper discussions and explanations—that is, to consider the why and how of the task. This ensured that the researcher's presence did not influence the children's behavior. Co-discovery provided a platform for the children to discuss and question every strategy in the game; this reflected not only what they were thinking but also why they were thinking it. This natural verbal communication between the children provided rich verbal data that represented the internal cognitive processes that were mapped into their interactions with the game system. They brought their own experiences and knowledge into the discussion. The two sets of knowledge and experiences provided a critical perspective on the understanding of the children's interactions.

The level of expertise of participants is crucial for pairing them in co-discovery (O'Malley, Draper, & Riley, 1984). Nielsen, Clemmensen, and Yssing (2002) favored pairing participants with the same level of expertise in co-discovery so that the expertise of any one participant did not influence the outcomes of the study. Kahler, Kensing, and Muller (2000), on the other hand, suggested pairing participants with different levels of expertise, thus enabling one person to guide the interaction. However, pairing children with different levels of experience could result in a child with higher levels of experience taking over the gameplay. Level of acquaintance is another important element that plays an important role in pairing participants in co-discovery (Als, Jensen, & Skov, 2005). Children often behave differently depending on how well they know each other. Rather than expertise, level of acquaintance was more important for this research study as it made children comfortable and encouraged natural behavior. Children who knew each other outside the research study—as friends, classmates, or siblings—were paired for the experiments in this research.

4.4.2 DIARIES

O'Brien (2010) used a diary study to collect the daily encounters of older people with new and infrequently used technologies over a period of 10 days. The timeframe was specifically chosen to include days of regular activities and weekend activities. Participants were provided with a daily journal where they recorded each new technology, new features/functions of standard technology, infrequently used technology, and infrequently used features/functions of standard technology. Participants entered the date and time of the encounter, details in regard to the availability of help, such as instructions and other people, and details of problems they had experienced in their interactions with any of the technologies, including those frequently used, everyday technologies reported on the inventory. Participants were not provided with any examples for the daily journal entry to avoid any probes into problem identification and definition. This method was able to uncover not only what technologies the participants used but also some of the motivations behind using them, their levels of knowledge about them and the problems they encountered with them, and how these differed between low- and high-tech older adults and younger adults.

4.4.3 RETROSPECTIVE INTERVIEWS

Structured or standardized interviews are useful if information from a number of respondents is to be combined and compared and can generate quantitative data, as the structure provides consistency (Sommer & Sommer, 1997). Interviews have been used by Blackler et al. (2010) and O'Brien (2010) to gain information about participants' familiarity with specific interface features after they have used an interface and have also been used as an alternative method of assessing familiarity without task completion or observation (Lawry et al., 2009, 2010). However, conscious events immediately fade in memory, and children in particular might not remember the details of gameplay during the interview (Popovic et al., 2012). In Desai, Blackler, and Popovic (2015, 2016), they were therefore shown the audio and video recordings of their play immediately after their play session. They talked about their play and about why and how they made their game decisions. They also asked their experiment partners questions about the game. Researcher intervention in terms of prompts or questions was required only when there was absolute silence; however, this was very rare.

4.4.4 SURVEYS AND QUESTIONNAIRES

As previously discussed, questionnaires of various designs have been used for much intuitive interaction research. The most commonly used would be the TF questionnaire and its derivatives, which have been used in many of the studies conducted at QUT (Cave et al., 2014; Desai et al., 2015; McEwan et al., 2014; Reddy et al., 2011) as well as adapted by O'Brien (2010) and Still and Still (2018). An early product of the IUUI research group was Evalint (*Eval*uate *int*uitive use), a questionnaire for evaluating intuitive interaction with prospective users of the product. The questionnaire consisted of four scales: perceived effortlessness of use, perceived error rate, perceived achievement of goals, and perceived effort of learning (Blackler & Hurtienne, 2007). Diefenbach and Ullrich (2015) also used a large survey that presented various scenarios to respondents to test the four components of their INTUI model. The QUESI covers a range of subjective measures of intuitive interactions and has been used by Hurtienne (2009) and Naumann and Hurtienne (2010) to test a number of products for intuitive interaction, and it has been applied by Still and Still (2018) in this volume. Many subjective measures and ratings are collected via questionnaires delivered after participants have completed tasks. For example, Fischer (2018) presented participants with surveys to get subjective feedback and McAran (2018) used well-established data collection and analysis methods commonly used within the TAM community but applied intuitive interaction theory to develop a new measure (Perceived Intuitiveness) and tested it using these established TAM methods via an online survey.

4.5 COMMERCIAL APPLICATIONS

Some researchers have developed or adapted intuitive interaction research methods for commercial applications, which has had some advantages and some disadvantages. The intensive feature-based methodology of Blackler et al. (2010) was adapted to a commercial project. We were asked to compare two models of the same product type for intuitiveness. Sixteen participants were divided into two groups of eight, with one group assigned to use each model. Each group was composed of a balanced selection of participants based on their age, educational background, TF score, and gender. All participants were video-recorded performing the same four tasks with their product.

Following the tasks, participants were asked to rate their familiarity with the features of the product interface and how difficult they found each of the tasks. Coding video of all 16 experiments, with four tasks in each, was obviously a time-consuming process when every feature use had to be coded, which was a disadvantage of this methodology for the purposes of commercial research. However, this approach did allow us to study the features of the two models very closely and to provide detailed feedback to the client on exactly how the interfaces compared and how each feature could be improved.

Fischer et al. (2015) report on an experiment in applying image schemas, but in this case they aimed to find a more efficient way of discovering and applying them, in order to improve the design process as well as the assessment of new interfaces. They focused on helping users to access existing schemas rather than having to construct new ones in order to save time and effort for both users and developers of interfaces. Also, Fischer (2018) reports on ways to apply intuitive interaction research to fast and agile product development using similar principles of understanding schema induction versus schema transfer and exploiting each to achieve the most efficient process.

Asikhia, Setchi, Hicks, and Walters (2015) offered a novel way to quantify intuitive interaction. They elicited image schemas from existing interfaces and used them to evaluate interfaces during usability testing. Their framework quantifies intuitive interaction by comparing the image schemas envisaged by the designer of a product with those used by its users. They evaluated the framework through a study involving participants completing a set task with a product. This study identified the image schemas that were correctly used in accordance with the designer's intention and those that were incorrectly used, which contributed to the difficulties that participants experienced. They proposed this approach as a framework for testing interfaces for their intuitiveness.

Palmer et al. (2018) undertook very well-used "quick and dirty" usability testing during the development of a health kiosk in Africa, with low participant numbers and standard measures such as NPR (Net Promoter Score), completion times, failure rates, and error counts while also comparing local and international user groups and their past experience of such products.

There is more to be done in the area as many research methods need streamlining to be suitable for use in industry. Chapters 9, 10, and 11 (Fischer, 2018; Hespanhol, 2018; Palmer et al., 2018) in particular include case studies that showcase some of the ways in which intuitive interaction theory has been applied and intuitive interaction has been assessed in real projects.

4.6 INTUITIVE INTERACTION RESEARCH METHODS TOOLKIT

The last 18 years of research has produced a firm understanding of intuitive interaction and how to apply intuitive interaction principles to design in the real world for a variety of user groups and applications. The development of robust and detailed methods has allowed these discoveries to occur, and teasing out these complex issues would not have been possible without such formalized methods. The methods used by us and others to investigate intuitive interaction have been various, but the underpinning approach has some kind of observation of people performing tasks with real or prototyped interfaces. Several other methods (e.g., surveys, diaries, questionnaires, interviews, verbal protocols, and the application of image schema theory) have been successfully used to supplement it, and automated software has sometimes been utilized to collect some of the performance data (e.g., Fischer et al., 2015; Hurtienne, 2009).

However, although some purely subjective measures have been used recently, it is unlikely that any other method could completely replace this approach as this is the only way researchers can see and understand what real people actually do with products and interfaces.

Table 4.1 provides a summary of the methods and tools used in intuitive interaction research. This Intuitive Interaction Research Methods Toolkit can be used by researchers and designers to identify the methods and techniques they can use to further their research on intuitive interaction and the design and development of intuitive products, systems, and interfaces. The toolkit consists of four areas: (1) data collection methods, (2) objectives, (3) goals, and (4) measures. For example, if the objective is to study the subjective experiences involved in intuitive interaction and the goal is to understand intuitive interaction as a process and to test interfaces and products for intuitive interaction, the appropriate data collection method is QUESI and the measures are *Subjective mental workload*, *Perceived achievement of goals*, *Effort of learning*, and *Familiarity*. Appropriate data collection methods and the measures to be used in analysis of the data can be accessed from the first column and the fourth column of Table 4.1, respectively. The main references provided will lead researchers and designers to the details of each method. Thus, if researchers are able to identify the goals and objectives of their intuitive interaction study, the methods that can be used in the study can be identified from the toolkit.

Each method group (column 1) is supported by the research on which it was built. The method groups included are observations, verbal protocol, questionnaires (QUESI, INTUI, TF), the TAM model, interviews, diaries, and the application of image schema theory. For each method group, goals, objectives, and measures are summarized. For example, if the main goal was an understanding of intuitive interaction (column 2, row 1), and the objective was to distinguish the intuitive process (column 3, row 1), the measures to achieve this are listed in column 4, row 1. The toolkit aims to cover both research into and the application of intuitive interaction. Thus, goals such as "Understanding intuitive interaction" may be more appropriate for researchers, whereas "Studying intuitive interaction with interfaces and products" may relate better to a commercial project. However, there are no hard-and-fast rules, and some goals (e.g., "Testing intuitive interaction design tools") could relate to either.

It should be noted that INTUI and QUESI questionnaires have the same goal and objective; however, the measures used in the associated citations are different. Researchers and designers should identify the measures that would be more appropriate for the context of their study, such as user groups, settings, and so on. On the other hand, observation and verbal protocol share objectives, measures, and citations as they are generally used together to achieve the same objectives and goals.

After nearly two decades of research, much has been done but much remains to be done, as outlined in Chapter 1 (Blackler, 2018). However, the robust methods developed have stood the test of time. They have been adapted and applied to differing participant groups across the world and across age ranges, have employed a variety of real and prototyped objects and interfaces, and have produced consistent and reliable results on which we can continue to build. These methods are integrated in the Intuitive Interaction Research Methods Toolkit (Table 4.1). The toolkit not

TABLE 4.1

Intuitive Interaction Research Methods Toolkit

Data Collection Methods	Goals	Objectives	Measures	Citation
Observations	Understanding intuitive interaction Studying intuitive interaction with interfaces and products Testing of intuitive interaction design tools	Distinguishing intuitive processing from other cognitive processes Quantifying performance measures	Time on task Click counts Latency Errors Coded intuitive uses	Blackler et al. (2010) O'Brien (2010) Brandenburg and Sachse (2012) McEwan et al. (2014) Reddy et al. (2011)
Verbal protocols (concurrent, retrospective, co-discovery)	Understanding intuitive interaction Studying intuitive interaction with interfaces and products Testing of intuitive interaction design tools		Correct uses Incorrect uses Accessing help Task recall	Lawry et al. (2011) Antle et al. (2009) Desai et al. (2015)
Questionnaire (QUESI)	Understanding intuitive interaction Studying intuitive interaction with interfaces and products	Measuring subjective consequences of intuitive interaction	Subjective mental workload Perceived achievement of goals Effort of learning Familiarity	Hurtienne (2009) Nauman and Hurtienne (2010)
Questionnaire (INTUI)	Understanding intuitive interaction Studying intuitive interaction with interfaces and products	Measuring subjective consequences of intuitive interaction	Gut feeling Magical experience Effortlessness Verbalizability Perceived intuitiveness	Ullrich and Diefenbach (2010 a, b) Blackler (2008a) Naumann and Hurtienne (2010)
TF questionnaires	Understanding intuitive interaction Understanding past experience Identifying design features for intuitive interaction	Quantifying TF	Exposure to interface features Frequency of use of interface features	Blackler et al. (2010) McEwan et al. (2014) Desai et al. (2015) Reddy et al. (2018)

(Continued)

TABLE 4.1 (CONTINUED)

Intuitive Interaction Research Methods Toolkit

Data Collection Methods	Goals	Objectives	Measures	Citation
Technology Acceptance Model (TAM)	Understanding intuitive interaction in the context of technology acceptance research	Quantifying perceived intuitiveness Measuring subjective consequences of intuitive interaction	Perceived ease of use Perceived usefulness Compatibility Perceived intuitiveness	McAran (2018)
Interviews	Understanding intuitive interaction Understanding past experience	Identifying Interaction with interfaces Qualifying TF	Individual features interaction/familiarity Overall task impressions or satisfaction	Blackler et al. (2010) O'Brien (2010) Desai et al. (2015, 2016)
Diaries	Understanding intuitive interaction Studying intuitive interaction with interfaces and products over time	Identifying technology experiences and TF	Technologies and features used Accessing help Correctness Task performance	O'Brien (2010)
Application of image schemas	Understanding intuitive interaction Studying intuitive interaction with interfaces and products Testing of intuitive interaction design tools	Identifying metaphorical extensions of image schemas Identifying metaphorical compatibility/incompatibility Identifying induction of new schemas/transfer of existing schemas Investigating embodied metaphors Linking images schemas to past experience	Accurate responses Response time Error rate Subjective judgment Goal specific tasks Decision times Completion times Existing image schemas	Hurtienne (2009) Hurtienne and Blessing (2007) Fischer et al. (2015) Fischer (2018) Asikhia et al. (2015) Antle et al. (2009)

only summarizes the methods used by various researchers in intuitive interaction all over the world but also provides direction for future design and research initiatives in intuitive interaction. It highlights the importance of identifying the objectives and goals of research and design in the area of intuitive interaction, as intuitive interaction has been studied by various researchers from different perspectives. Thus, it is important that future research and design work on intuitive interaction has clear goals and objectives, so appropriate methods are selected.

4.7   CONCLUSION

This research area has the potential to help make products and systems intuitive to use. However, in order to investigate complex and unusual issues like this one, detailed and robust methods that can be tailored to suit each situation are needed. This chapter has demonstrated the emergence of well-supported methods to research and apply intuitive interaction to products and systems. The data collection methods are common. However, when applied, they become domain-specific methods, unified by their goals, objectives, coding schemes, and measures. This is evident where different researchers have applied similar approaches to understanding the intuitive interaction domain. The Intuitive Interaction Research Methods Toolkit (Table 4.1) is designed to provide a summary of tried-and-tested intuitive interaction research methods. Its strength is to identify the approaches that will be most effective at bringing together intuitive interaction research and applications.

REFERENCES

Adebesin, T., De Villiers, M., & Ssemugabi, S. (2009). Usability testing of e-learning: An approach incorporating co-discovery and think-aloud. Paper presented at the 2009 Annual Conference of the Southern African Computer Lecturers' Association, Eastern Cape, South Africa.

Agor, W. H. (1986). *The Logic of Intuitive Decision Making: A Research-Based Approach for Top Management*. Quorum Books, New York, NY.

Als, B. S., Jensen, J. J., & Skov, M. B. (2005). Comparison of think-aloud and constructive interaction in usability testing with children. Paper presented at the 2005 Conference on Interaction Design and Children, Boulder, CO.

Antle, A. N., Corness, G., & Droumeva, M. (2009). What the body knows: Exploring the benefits of embodied metaphors in hybrid physical digital environments. *Interacting with Computers*, 21(1–2), 66–75.

Asikhia, O. K., Setchi, R., Hicks, Y., & Walters, A. (2015). Conceptual framework for evaluating intuitive interaction based on image schemas. *Interacting with Computers*, 27(3), 287–310.

Bastick, T. (2003). *Intuition: Evaluating the Construct and Its Impact on Creative Thinking*. Kingston, Jamaica: Stoneman and Lang.

Berry, D. C., & Broadbent, D. E. (1988). Interactive tasks and the implicit–explicit distinction. *British Journal of Psychology*, 79(2), 251–272.

Bignami-Van Assche, S., Reniers, G., & Weinreb, A. A. (2003). An assessment of the KDICP and MDICP data quality: Interviewer effects, question reliability and sample attrition. *Demographic Research*, 1, 31–76.

Blackler, A. (2008a). *Intuitive Interaction with Complex Artefacts: Empirically-Based Research*. Saarbrücken, Germany: VDM.

Blackler, A. (2008b). Applications of high and low fidelity prototypes in researching intuitive interaction. Paper presented at the 2008 Design Research Society Conference, "Undisciplined!," Sheffield, UK.

Blackler, A. (2018). Intuitive interaction: An overview. In A. Blackler (Ed.), *Intuitive Interaction: Research and Application* (pp. 3–18). Boca Raton, FL: CRC Press.

Blackler, A., & Hurtienne, J. (2007). Towards a unified view of intuitive interaction: Definitions, models and tools across the world. *MMI-Interaktiv, 13*(August), 37–55.

Blackler, A., Popovic, V., & Mahar, D. (2003). The nature of intuitive use of products: An experimental approach. *Design Studies, 24*(6), 491–506.

Blackler, A., Popovic, V., & Mahar, D. (2010). Investigating users' intuitive interaction with complex artefacts. *Applied Ergonomics, 41*(1), 72–92. doi: 10.1016/j.apergo.2009.04.010

Blackler, A., Popovic, V., Lawry, S., Reddy, R. G., Doug Mahar, Kraal, B., & Chamorro-Koc, M. (2011). Researching intuitive interaction. Paper presented at IASDR 2011, Delft, the Netherlands.

Blackler, A., Popovic, V., & Mahar, D. (2014). Applying and testing design for intuitive interaction. *International Journal of Design Sciences and Technology, 20*(1), 7–26.

Blackler, A., Popovic, V., Mahar, D., Reddy, R. G., & Lawry, S. (2012). Intuitive interaction and older people. Paper presented at DRS 2012, "Research: Uncertainty, Contradiction and Value," Bangkok, Thailand.

Bowers, K. S., Regehr, G., Balthazard, C., & Parker, K. (1990). Intuition in the context of discovery. *Cognitive Psychology, 22*(1), 72–110.

Brandenburg, S., & Sachse, K. (2012). Intuition comes with experience. Paper presented at the Human Factors and Ergonomics Society of Europe Conference, "Human Factors: A View from an Integrative Perspective." Toulouse, France.

Brooke, J. (1996). SUS: A quick and dirty usability scale. *Usability Evaluation in Industry, 189*(194), 4–7.

Cave, A., Blackler, A. L., Popovic, V., & Kraal, B. J. (2014). Examining intuitive navigation in airports. Paper presented at DRS 2014, "Design Big Debates: Pushing the Boundaries of Design Research," Umea, Sweden.

Chattopadhyay, D., & Bolchini, D. (2015). Motor-intuitive interactions based on image schemas: Aligning touchless interaction primitives with human sensorimotor abilities. *Interacting with Computers, 27*, 327–343.

Cockayne, A., Wright, P. C., & Fields, B. (1999). Supporting interaction strategies through the externalization of strategy concepts. Paper presented at INTERACT, Edinburgh, UK.

Dell, N., Vaidyanathan, V., Medhi, I., Cutrell, E., & Thies, W. (2012). Yours is better! Participant response bias in HCI. Paper presented at the SIGCHI Conference on Human Factors in Computing Systems, Austin, TX.

Desai, S., Blackler, A., & Popovic, V. (2015). Intuitive use of tangible toys. Paper presented at IASDR 2015, "Interplay," Brisbane, Australia.

Desai, S., Blackler, A., & Popovic, V. (2016). Intuitive interaction in a mixed reality system. Paper presented at DRS 2016, Brighton, UK.

Diefenbach, S., & Ullrich, D. (2015). An experience perspective on intuitive interaction: Central components and the special effect of domain transfer distance. *Interacting with Computers, 27*(3), 210–234.

Dreyfus, H. L., Dreyfus, S. E., & Athanasiou, T. (1986). *Mind Over Machine: The Power of Human Intuition and Expertise in the Era of the Computer.* New York: Free Press.

Engeström, Y., Miettinen, R., & Punamäki, R.-L. (1999). *Perspectives on activity theory.* Cambridge University Press, Cambridge, UK.

Fischbein, E. (1987). *Intuition in Science and Mathematics.* Dordrecht, the Netherlands: Reidel.

Fischer, S. (2018). Designing intuitive products in an agile world. In A. Blackler (Ed.), *Intuitive Interaction: Research and Application* (pp. 195–212). Boca Raton, FL: CRC Press.

Fischer, S., Itoh, M., & Inagaki, T. (2015). Screening prototype features in terms of intuitive use: Design considerations and proof of concept. *Interacting with Computers*, 27(3), 256–270.

Gilhooly, K., Fioratou, E., & Henretty, N. (2010). Verbalization and problem solving: Insight and spatial factors. *British Journal of Psychology*, 101(1), 81–93.

Gudur, R. R., Blackler, A., Popovic, V., & Mahar, D. (2009). Redundancy in interface design and its impact on intuitive use of a product in older users. Paper presented at IASDR 2009, Seoul, South Korea.

Gudur, R. R., Blackler, A. L., Popovic, V., & Mahar, D. (2013). Ageing, technology anxiety and intuitive use of complex interfaces. *Lecture Notes in Computer Science: Human–Computer Interaction; INTERACT 2013, 8119*, 564–581.

Hammond, K. R. (1993). Naturalistic decision making from a Brunswikian viewpoint: Its past, present, future. In G. A. Klein, J. Orasanu, R. Calderwood & C. E. Zsambok (Eds.), *Decision Making in Action: Models and Methods* (pp. 205–227). Norwood, NJ: Ablex.

Hart, S. G., & Staveland, L. E. (1988). Development of NASA-TLX (Task Load Index): Results of empirical and theoretical research. *Advances in Psychology*, 52, 139–183.

Hassenzahl, M., Burmester, M., & Koller, F. (2008). Der User Experience (UX) auf der Spur: Zum Einsatz von www.attrakdiff.de. Paper presented at the Usability Professionals conference, Stuttgart, Germany.

Hawthorn, D. (2007). Interface design and engagement with older people. *Behaviour & Information Technology*, 26(4), 333–341.

Hertzum, M. (2016). Usability testing: Too early? Too much talking? Too many problems? *Journal of Usability Studies*, 11(3), 83–88.

Hespanhol, L. (2018). City context, digital content and the design of intuitive urban interfaces. In A. Blackler (Ed.), *Intuitive Interaction: Research and Application* (pp. 173–194). Boca Raton, FL: CRC Press.

Höysniemi, J., Hämäläinen, P., & Turkki, L. (2003). Using peer tutoring in evaluating the usability of a physically interactive computer game with children. *Interacting with Computers*, 15(2), 203–225.

Hurtienne, J. (2009). Image schemas and design for intuitive use. PhD thesis, Technischen Universität Berlin, Germany.

Hurtienne, J., & Blessing, L. (2007). Design for intuitive use: Testing image schema theory for user interface design. Paper presented at the 16th International Conference on Engineering Design, Paris, France, 2007.

Hurtienne, J., Horn, A. M., & Langdon, P. M. (2010). Facets of prior experience and their impact on product usability for older users. In P. M. Langdon, P. J. Clarkson & P. Robinson (Eds.), *Designing Inclusive Interactions* (pp. 123–132). London: Springer.

Hurtienne, J., Klöckner, K., Diefenbach, S., Nass, C., & Maier, A. (2015). Designing with image schemas: Resolving the tension between innovation, inclusion and intuitive use. *Interacting with Computers*, 27, 235–255.

Kahler, H., Kensing, F., & Muller, M. (2000). Methods & tools: Constructive interaction and collaborative work; Introducing a method for testing collaborative systems. *Interactions*, 7(3), 27–34.

King, L., & Clark, J. M. (2002). Intuition and the development of expertise in surgical ward and intensive care nurses. *Journal of advanced nursing*, 37(4), 322–329.

Klein, G. (1998). *Sources of Power: How People Make Decisions*. Cambridge, MA: MIT Press.

Lawry, S., Popovic, V., & Blackler, A. (2009). Investigating familiarity in older adults to facilitate intuitive interaction. Paper presented at IASDR 2009, Seoul, South Korea.

Lawry, S., Popovic, V., & Blackler, A. (2010). Identifying familiarity in older and younger adults. Paper presented at the Design and Complexity Design Research Society International Conference 2010, Montreal, Canada.

Lawry, S., Popovic, V., & Blackler, A. (2011). Diversity in product familiarity across younger and older adults. Paper presented at IASDR 2011, Delft, the Netherlands.

Lewis, T., Langdon, P. M., & Clarkson, P. J. (2008). Prior experience of domestic microwave cooker interfaces: A user study. In P., Langdon, J., Clarkson, and P., Robinson (Ed.) *Designing Inclusive Futures* (pp. 3–14). Springer, London, UK.

Lim, K. H., Ward, L. M., & Benbasat, I. (1997). An empirical study of computer system learning: Comparison of co-discovery and self-discovery methods. *Information Systems Research, 8*(3), 254–272.

Macaranas, A., Antle, A. N., & Riecke, B. E. (2015). Intuitive interaction: Balancing users' performance and satisfaction with natural user interfaces. *Interacting with Computers, 27*(3), 357–370.

Mayo, C. W., & Crockett, W. H. (1964). Cognitive complexity and primacy–recency effects in impression formation. *The Journal of Abnormal and Social Psychology, 68*(3), 335.

McAran, D. (2018). Development of the technology acceptance intuitive interaction model. In A. Blackler (Ed.), *Intuitive Interaction: Research and Application* (pp. 129–150). Boca Raton, FL: CRC Press.

McClelland, I. (1995). Product assessment and user trials. In J. R. Wilson & E. N. Corlett (Eds.), *Evaluation of Human Work: A Practical Ergonomics Methodology* (2nd ed., pp. 249–284). London: Taylor and Francis.

McEwan, M., Blackler, A., Johnson, D., & Wyeth, P. (2014). Natural mapping and intuitive interaction in videogames. Paper presented at CHI PLAY '14, First ACM SIGCHI Annual Symposium on Computer–Human Interaction in Play, Toronto, Canada.

Mihajlov, M., Law, E. L.-C., & Springett, M. (2014). Intuitive learnability of touch gestures for technology-naïve older adults. *Interacting with Computers, 27*(3), 344–356.

Mohan, G., Blackler, A. L., & Popovic, V. (2015). Using conceptual tool for intuitive interaction to design intuitive website for SME in India: A case study. Paper presented at IASDR 2015, Brisbane, Australia.

Mohs, C., Hurtienne, J., Israel, J. H., Naumann, A., Kindsmüller, M. C., Meyer, H. A., & Pohlmeyer, A. (2006). IUUI: Intuitive Use of User Interfaces. Paper presented at Usability Professionals 2006, Stuttgart, Germany.

Nardi, B. A. (1996). *Context and Consciousness: Activity Theory and Human–Computer Interaction*. Cambridge, MA: MIT Press.

Naumann, A. Hurtienne, J., & (2010). *Benchmarks for Intuitive Interaction with Mobile Devices*. Mobile HCI'10 September 7–10, Lisboa, Portugal. ACM 978-1-60558-835-3.

Naumann, A., & Hurtienne, J. (2010). Benchmarks for intuitive interaction with mobile devices. Paper presented at the 12th International Conference on Human–Computer Interaction with Mobile Devices and Services, Lisboa, Portugal.

Naumann, A., Hurtienne, J., Israel, J. H., Mohs, C., Kindsmüller, M. C., Meyer, H. A., & Hußlein, S. (2007). Intuitive use of user interfaces: Defining a vague concept. Paper presented at the International Conference on Engineering Psychology and Cognitive Ergonomics, Beijing, China.

Nielsen, J., Clemmensen, T., & Yssing, C. (2002). Getting access to what goes on in people's heads? Reflections on the think-aloud technique. Paper presented at the Second Nordic Conference on Human–Computer interaction, Aarhus, Denmark.

Noddings, N., & Shore, P. J. (1984). *Awakening the Inner Eye: Intuition in Education*. ERIC, Teachers College Press, New York, NY.

O'Brien, M. (2018). Lessons on intuitive usage from everyday technology interactions among younger and older people. In A. Blackler (Ed.), *Intuitive Interaction: Research and Application* (pp. 89–112). Boca Raton, FL: CRC Press.

O'Brien, M. A. (2010). *Understanding Human–Technology Interactions: The Role of Prior Experience and Age*. Atlanta: Georgia Institute of Technology.

O'Malley, C. E., Draper, S. W., & Riley, M. S. (1984). Constructive interaction: A method for studying human–computer–human interaction. Paper presented at IFIP INTERACT 1984, London, UK.

Palmer, J., Ogunyoka, T., & Hammond, C. (2018). Intuitive interaction in industry user research: Context is everything. In A. Blackler (Ed.), *Intuitive Interaction: Research and Application* (pp. 213–226). Boca Raton, FL: CRC Press.

Popovic, V. (2003). An approach to knowledge generation by research and its utilisation in practice: Situating doctoral research around the artifacts. In D., Durling and K., Sugiyama (Eds.) 3rd Doctoral Education in Design, Tsukuba, Japan.

Popovic, V., Kraal, B. J., Blackler, A. L., & Chamorro-Koc, M. (2012). Observational research and verbal protocol methods. In *Proceedings of the Design Research Society (DRS) 2012*, 311–324.

Reddy, G. R., Blackler, A., Popovic, V., & Mahar, D. (2011). Ageing and use of complex product interfaces. Paper presented at IASDR 2011, "Diversity and Unity," Delft, the Netherlands.

Reddy, R. G., Blackler, A., Popovic, V., & Mahar, D. (2009). Redundancy in interface design and its impact on intuitive use of a product in older users. Paper presented at IASDR 2009, Seoul, South Korea.

Reddy, R. G., Blackler, A., & Popovic, V. (2018). Adaptable Interface Framework for Intuitively Learnable Product Interfaces for People With Diverse Capabilities. In A. Blackler (Ed.), *Intuitive Interaction: Research and Application* (pp. XX). Boca Raton, FL: CRC Press.

Reinhardt, D., & Hurtienne, J., (2017). Interaction Under Pressure: Increased Mental Worklaod Makes Issues of Intuitve Interaction Visible. In *Proc ACM Designing Interactive Systems*, Edinburgh, UK, 67–71.

Ryan, R. (2006). Intrinsic motivation inventory (IMI). Self-Determination Theory: An Approach to Human Motivation and Personality. Online: http://www.psych.rochester.edu/SDT/measures/IMI_description.php (accessed October 9, 2009).

Salk, J. (1983). *Anatomy of Reality*. New York, NY: Columbia University Press.

Satmetrix Systems, Bain & Company, & Reichheld, F. (2017). What is net promoter? *Harvard Business Review*. Accessed 23/3/18, www.netpromoter.com/know.

Sommer, B. B., & Sommer, R. (1997). *A Practical Guide to Behavioral Research: Tools and Techniques* (4th ed). New York: Oxford University Press.

Still, J. D., & Still, M. L. (2018). Cognitively describing intuitive interactions. In A. Blackler (Ed.), *Intuitive Interaction: Research and Application* (pp. 41–62). Boca Raton, FL: CRC Press.

Still, J. D., Still, M. L., & Grgic, J. (2015). Designing intuitive interactions: Exploring performance and reflection measures. *Interacting with Computers*, 27(3), 271–286.

Tretter, S., Diefenbach, S., & Ullrich, D. (2018). Intuitive interaction from an experiential perspective: The intuitivity illusion and other phenomena. In A. Blackler (Ed.), *Intuitive Interaction: Research and Application* (pp. 151–170). Boca Raton, FL: CRC Press.

Ullrich, D., & Diefenbach, S. (2010a). From magical experience to effortlessness: An exploration of the components of intuitive interaction. Paper presented at the Sixth Nordic Conference on Human–Computer Interaction: "Extending Boundaries," Reykjavik, Iceland.

Ullrich, D., & Diefenbach, S. (2010b). INTUI: Exploring the facets of intuitive interaction. In J., Ziegler and A., Schmidt (Eds.), *Mensch und Computer 2010: Interaktive Kulturen*, Oldenbourg Publishing House, Berlin, Germany. *10*, 251–260.

Vermeeren, A. (1999). Designing scenarios and tasks for user trials of home electronic devices. In W. S. Green & P. W. Jordan (Eds.), *Human Factors in Product Design: Current Practice and Future Trends* (pp. 47–55). London: Taylor and Francis.

Xie, L., Antle, A. N., & Motamedi, N. (2008). Are tangibles more fun? Comparing children's enjoyment and engagement using physical, graphical and tangible user interfaces. Paper presented at the Second International Conference on Tangible and Embedded Interaction, Bonn, Germany.

5 Lessons on Intuitive Usage from Everyday Technology Interactions among Younger and Older People

Marita A. O'Brien

CONTENTS

5.1 INTRODUCTION

If past experience informs intuitive interaction with technologies, then one would expect that most older people who have used a variety of everyday technologies such as household appliances and entertainment systems throughout their lives would be able to use new technologies to achieve tasks in these domains with little difficulty. Yet, a recent Pew Research survey reported that 73% of older people (aged 65 and older) prefer or strongly prefer someone to help them set up or first use a new tech-nology (Anderson & Perrin, 2017). This preference decreases as age decreases, but even so, 62% of the next cohort of older people (now aged 50–65) also indicated a preference for external support. These findings suggest that knowledge from past experience is insufficient for successful first use of modern technologies. Chapter 1 of this volume (Blackler, 2018) summarizes research suggesting that designing for familiarity increases intuitive use among older people, but the author also cautions that this familiarity "does not always work as expected" (pp. 3–18). Thus, there is a significant need to understand how older people use familiarity and prior knowledge during technology interactions to improve predictions about effective designs.

This chapter describes research that investigated how age differences and knowl-edge differences affect the use of everyday technologies, which are technologies used for ordinary tasks among people in their typical surroundings (O'Brien & Rogers, 2013). The definition of intuitive interaction used in this research was "interactions between humans and high technology in lenient learning environments that allow the human to use a combination of prior experience and feedforward methods to achieve their functional and abstract goals" (O'Brien, Rogers, & Fisk, 2010, p. 107). Because users are presumed to have relevant task knowledge and skills, the technologies are typically designed for use with little training or instruction. Note that even though everyday technologies can be used intuitively according to the definitions proposed in Chapter 1 of this book, prior research (e.g., Blackler, Popovic, & Mahar, 2010) has reported some users instead use other approaches while interacting with the same technology, such as conscious reasoning. This conscious reasoning approach is slower and more effortful than intuitive usage, but it led numerically to a higher per-centage of correct uses in Blackler, Popovic, & Mahar's (2010) investigation of par-ticipant interactions with a digital camera. Because the intuitive interaction approach was still more successful overall when time and "correct but inappropriate" uses were also considered, the presence of this reasoned approach by some participants suggests that it may have other advantages, such as higher perceived control for users to check their work. Indeed, several researchers (Hammond, Hamm, Grassia, & Pearson, 1987; Kahneman & Frederick, 2002) have proposed that humans implic-itly determine whether to use their faster, intuitive cognitive processes or slower, reflective cognitive processes for individual behaviors based on user, system, and environmental attributes. These attributes include stimuli characteristics, error con-sequences, and the user's certainty in the task process, which represent both external and internal sources of knowledge that could elicit intuitive or reflective interaction approaches. Thus, considering how user knowledge and technology familiarity con-tribute to successes and problems with these technologies could inform the design of pathways to intuitive use, facilitating the design of technologies that are intuitive to use for a diverse population.

Age differences in the studies discussed here were identified in an extreme groups design that compared high-tech older people with college students aged 18–28. Because both high-tech older people and these college students had similar recent technology experience levels, problems experienced only by the high-tech older people could be due to differences in technology generation or to typical age-related differences such as perceptual decline. On the other hand, one might expect older people to be more successful at using classic household technologies because they have had more experience with a variety of instances of that technology over the years.

The general approach for examining knowledge differences in these studies was an extreme groups design that compared older people in the same age range (65–75) who differed only in recent technology experience (low or high tech) but were from the same technology generation. Docampo Rama (2001) coined the term *technology generation* to describe an age cohort with experience using similar technologies before age 25. Because these technologies have the same operations, interaction techniques, and procedures, users can rapidly access this knowledge similarly to a native language. For example, the current cohort of older people (including those aged at least 57 in 2018) matured when the electromechanical interaction style featuring 1:1 mapping between a control and function predominated. Docampo Rama proposed that individuals from one generation can use technologies from a different generation, but this non-native use takes more effort.

To propose recommendations for increasing intuitive use in a broader population, relevant findings from both a naturalistic diary and a laboratory study using this extreme groups paradigm are described and discussed in this chapter. Quantitative analysis was performed in both studies to identify group differences, and qualitative analysis of detailed verbal reports from both studies was performed to understand the actual basis for differential use. Further details of both studies can be found in O'Brien (2010).

5.2 NATURALISTIC TECHNOLOGY USE

5.2.1 METHOD

Our first step was to understand the knowledge that is deployed among typical people in the populations of interest. Thus, we conducted a diary study that examined technology use over a 10-day period among ten college students and twenty older people from the community (O'Brien, Rogers, & Fisk, 2012). The older people represented an equal number of individuals with technology experience in the top third and bottom third of their cohort, as determined by a survey of over 600 older and younger people in the southeastern United States (Olson, O'Brien, Rogers, & Charness, 2011). After daily reporting of technology interactions for 10 days, participants returned to the lab to elaborate on their use of each technology and to describe successes and problems experienced during this timeframe. Each technology encounter was coded to identify the reason for the success or problem, focusing particularly on sources of participant knowledge. Internal sources of knowledge included prior experience and past training that built user knowledge and expectancies for the interaction. External sources of knowledge included technology labels, device feedback, instruction manuals, and other people present in the immediate environment. Reason codes also identified participant strategies for resolving problems and achieving their goals in spite of knowledge gaps and design flaws. For example, a combined knowledge code indicated that participants used both internal and external knowledge to complete the task. Chi-square analysis using $p = .05$ was then conducted to quantitatively identify group differences in the total number of successes and problems and the primary reasons for these results.

5.2.2 RESULTS

Our overall finding from this study (O'Brien et al., 2012) was that participants' technology experience was not the only factor determining the success of technology encounters but that a more diverse description of relevant knowledge was also needed. Familiarity has been defined as a measure of how much is recognized and understood about a given context as a result of prior knowledge. Experience is the activities that led to that knowledge being acquired. Hence, two individuals could undergo the same experience, but the knowledge they gain from that experience may be different (Lawry, 2012). As will be described in this section, recent technology experience discriminated the breadth and depth of technology-specific knowledge, but it did not discriminate types of problems experienced by group participants. Functional knowledge, such as task goals and procedures, also provided critical information for successful use and problem solving in every group. Identifying participants' metacognitive knowledge, such as their computer self-efficacy and primary technology interaction strategies, was also necessary to explain group differences.

5.2.2.1 Role of Prior Experience

The first general source of knowledge was general technology experience, assessed in the survey used to identify high-tech and low-tech older people for this study. This experience not only indicates familiar technologies but also specific features and controls, interaction techniques, expected guidance and feedback, and typical problem-solving techniques for these technologies. This experience was represented in similar technology repertoires for younger people and high-tech older people, although the older people used more kitchen and home health-care devices. These older people used devices with a wide range of ages, including gadgets received for their weddings decades ago, but younger people also occasionally used older technologies such as microfilm readers. Both groups used a range of functions on both cell phones and PCs, with participants in both groups discovering new features during the diary study period. In contrast, low-tech older participants used a broad set of technologies but significantly fewer PC and cell phone functions ($\chi^2[1, n=40]=22.5$, $p<.001$) (O'Brien et al., 2012). Discussions about their technology use suggested that they could even use novel technologies if they were interested in completing specific tasks that required them. In general, though, the low-tech older people clearly preferred familiar devices and functions and described specific efforts to obtain in-person assistance completing tasks even if the technology option might be more convenient. Thus, the technology experience score discriminated technology attitudes as well as technology experience and knowledge.

Although prior laboratory investigations have shown that older people have more problems in their technology interactions than younger people (e.g., Freudenthal & Mook, 2003; Kang & Yoon, 2008), neither high-tech nor low-tech older people reported more problems in our diary study. Specifically, younger people and high-tech older people had a similar number of problems, though for different reasons (O'Brien, Weger, DeFour, & Reeves, 2011). Younger people were more likely to report that the technology design conformed to their prior knowledge, but problems typically arose for three reasons: (1) interference from another specific technology, (2) expectations of different guidance or feedback, or (3) expectations of more time familiarizing themselves with a new version of the technology. In contrast, high-tech older people more generally reported that device guidance was insufficient. Several specifically blamed their declining memory for their lack of relevant knowledge, though high-tech older participants could typically describe technology guidance and feedback quite well. Perhaps because high-tech older people had sufficient background knowledge to interpret feedback, they could and indeed preferred to use instruction manuals to complement their knowledge for successful technology use in the first place.

Low-tech older people used fewer technologies overall and experienced fewer problems than high-tech older people. For both successes and problems during technology encounters, low-tech older people described technology guidance and feedback quite vaguely. This lower recall of feedback may stem from lower attention to guidance and feedback during the encounter, as might be expected from individuals who prefer to interact with people over technologies. The consequence of this lower attentional allocation is less error monitoring and lower awareness of seldom-used functions, making problem resolution more difficult and thus reinforcing low-tech participants' preference for well-known technologies (e.g., Langdon, Lewis, & Clarkson, 2007; Was & Woltz, 2013).

5.2.2.2   Role of Functional Knowledge

The second source of knowledge is functional knowledge related to the domain or task being completed, including knowledge of common functions, prerequisites, and the general sequence of procedural steps and milestones to successfully achieve task goals (O'Brien et al., 2012). These functional elements were usually noted as participants described their successes because they helped individuals know whether they were correctly making progress. Even when participants described using a new version of a technology such as a microwave, they were likely to note that they were successful because of functional knowledge such as "The corn was popped." Participants playing games on new technologies also described goals of competition and enjoyment that were facilitated by helpful friends providing encouragement and minor corrections to their interactions. Thus, participants seemed able to focus on achieving their goals because the technology design was consistent with their mental representation of the device's features and operational processes. This finding is consistent with prior research noting that generally correct mental representations allow users to resolve problems encountered with everyday technologies (Payne, 2003). Because these users achieve their goals with little attention to the technology itself, they may be quite satisfied with the technology and unlikely to recall problems that would be noticed if participants were completing tasks in a lab. Of critical importance for designers is the indication that participants' success was facilitated by internal knowledge rather than external knowledge from device features and feedback.

5.2.2.3   Role of Metacognitive Knowledge

The third type of relevant knowledge for successful everyday technology interactions is metacognitive knowledge, which is the knowledge and beliefs about one's self and others necessary to achieve cognitive tasks (Flavell, 1979). For example, participants in all our study groups stated particular types of self-efficacy that helped (e.g., computer self-efficacy) or hindered (e.g., memory self-efficacy) successful interactions. Low-memory self-efficacy led several participants to avoid multistep interactions without a manual or an expert nearby. Participants in all groups also noted contextual factors such as the consequence of errors, the presence of others, or device ownership that influenced how they interacted with the technology in the first place or tried to resolve problems.

Generalizing from individual encounters in different contexts, both younger people and high-tech older people seemed to recognize that occasional problems when interacting with new technologies are to be expected but could usually be worked through provided that error consequences were low. For example, one younger participant described playing a game on the first iPhone he had ever used. He had to experiment with the touches and gestures and learn to expect screen orientation changes when he rotated the device. Feelings of familiarity were the primary basis for users deciding whether to use this "trial-and-error" strategy or to appropriately select from their repertoire of well-known strategies for handling more complex tasks, such as manuals, internet-based instructional videos, or other people. Note that the trial-and-error strategy described in this chapter includes the fast and effortless characteristics of intuitive cognition described in Section 5.1, whereas the strategy of consulting support during first use includes the slower pace and conscious processing characteristics of reflective cognition also described in Section 5.1. The understanding that using this support strategy up front may take more time but will save frustration when choosing among unfamiliar options exemplifies relevant metacognitive knowledge that facilitates successful interaction with new technologies.

For a user with high technology self-efficacy, familiarity may also suggest that they have sufficient experience with similar products for relevant knowledge to trigger ideas that can be verified as they interact, even if they cannot report the source of prior knowledge. Provided that device feedback allows the user to quickly assess whether an action worked or to effortlessly select an alternate action, this trial-and-error strategy may be characterized by the user as intuitive because they are not fully conscious of the reasons for choosing each action. Indeed, the fact that feedback assessment and next action selection was quick and effortless may make the serial nature of their interaction seem like the actions occur in parallel. More importantly, as Still and Still (2018) describe in Chapter 3, these characteristics of implicit knowledge are fundamental to building users' knowledge of conventions that can be tapped for intuitive design. Designers could thus expect to leverage feature familiarity and other aspects of intuitive processing such as rapid knowledge access through associations (Kahneman & Frederick, 2002; Sloman, 1996) to elicit intuitive interactions for high-tech populations.

In contrast, low-tech older people rarely described intuitive usage during technology encounters. Instead, low-tech older people described a general strategy for completing higher-risk tasks, even with familiar technologies, of ensuring that they were alone and not rushed so that they could carefully focus on their task without distraction. For example, one low-tech participant always used recipes that were exactly followed step by step for her daily task of cooking dinner. This self-conscious and effortful approach represents a reflective strategy driven by user preference rather than inadequate prior experience. The fact that familiarity did not elicit intuitive processing for this group suggests the importance of understanding the metacognitive characteristics of target user groups for more accurate prediction of design impacts.

Although these metacognitive strategies of high- and low-tech people may seem quite different, the cases suggest strong user preferences to avoid errors among all participants. Younger people noted specific examples of problems with specific technologies that they avoided with known workarounds, but older people were more likely to just recall that there were problems with a specific technology that reduced their confidence to successfully use that device without external support. As one high-tech older adult described problems finding a cruise by surfing the web, he admitted typically employing a travel agent for these tasks because he frequently gets lost on the internet. This "lostness," when they cannot return to their intended path, can thus lead even typically high-tech older people to avoid certain technologies (Barnard, Bradley, Hodgson, & Lloyd, 2013). Alternatively, they learn to invest time up front, as another high-tech older adult noted when completing the infrequent task of replacing a toner cartridge on her printer, by reading the instructions completely and following them step by step (O'Brien & Rogers, 2013). Together with the other types of knowledge identified, these findings suggest that users have developed approaches to achieving their goals in spite of insufficient specific technology knowledge. New technology designs that facilitate continued use and the development of these first-use approaches are more likely to be perceived as familiar and thus more likely to be adopted.

5.3 LABORATORY EXAMINATION OF CLASSIC TECHNOLOGY USE

With the understanding of relevant knowledge for everyday technology encounters gained through the diary study (O'Brien et al., 2012), we could now examine specifically how participants from these three cohorts completed typical tasks on the same three devices in a laboratory setting similar to comparable studies (Blackler, Popovic, & Mahar, 2010; Freudenthal & Mook, 2003; Kang & Yoon, 2008). Video-recorded observations were made during the interactions, and participants were then interviewed to document specific prior knowledge that may have been accessed for each device feature, similar to the procedure used by Blackler, Popovic, & Mahar (2010). Lastly, participants were debriefed to identify their perceived cognitive workload, success, and satisfaction with the interaction (O'Brien, 2010).

This chapter only describes interaction with a *classic* device, by which we mean that the functions are standard and well known within the culture and across the technology generations being studied (O'Brien, 2010). We chose a radio alarm clock as this classic technology because clock radios have been widely available in the United States since at least the 1950s. Alarm clocks have evolved new features since their introduction, but base functions have remained essentially the same. In our

diary study, we noted that at least four participants in each of the three groups frequently used alarm clocks and that at least one older participant in both the high- and
low-tech groups reset the time on multiple alarm clocks during the study period.
This extensive usage by multiple participants across age and technology experience
suggested that all participants owned at least one clock radio and that the use of different clock radios is a familiar technology interaction for all groups. Thus, it seemed
likely that many older participants would have used several different models of an
alarm clock across different technology generations. Older people's longer lifespans,
meaning they had used a variety of clock radios with different controls and procedures, was expected to facilitate their interaction more than younger people because
varied interactions with different versions of a device can facilitate the acquisition of
a more accurate mental model for the type of device (Chen & Mittel, 2006). On the
other hand, age-related declines in cognitive and perceptual abilities may alter older
participant interactions even if the participant has the same familiarity, recency, and
frequency of use of the base technology (Blackler, Mahar, & Popovic, 2010; Reddy,
Blackler, Mahar, & Popovic, 2010).

5.3.1 METHOD

Twelve new participants were selected for each group (i.e., younger people, high-
tech older people, low-tech older people) using the same criteria as described for
the naturalistic diary study. As shown in Table 5.1, expectations about participants'
substantial experience with alarm clocks were confirmed. The wider experience for
older participants (more than five clock radios in their lifetime) was also confirmed,
though more than half of the younger participants had used at least two clock radios
previously. No significant age or experience differences between breadth, recency,
or frequency of alarm clock use were identified. Note that one or two participants in
each group had used this specific alarm clock before, but none owned it; later analysis of these participant interactions indicated that they were similar to their cohort's.

The specific clock radio selected for this study is shown in Figure 5.1. This clock
was selected to appeal to younger people because of the modern Sony brand name

FIGURE 5.1   Sony alarm clock radio used for exemplar technologies study (author's
photograph).

TABLE 5.1

Self-Reported Prior Experience with Clock Radios

	Younger People (n = 12)		High-Tech Older People (n = 12)		Low-Tech Older People (n = 12)	
	M	**SD**	**M**	**SD**	**M**	**SD**
Have you ever used an alarm clock before? (% yes)	100.0%	–	91.7%[a]	–	100.0%	–
If yes, approximately how many alarm clocks have you used in your lifetime? (% used)						
1	25.0%		0%		8.3%	
2–5	33.3%		16.7%		25.0%	
More than 5	41.7%		75.0%		66.7%	
If you use an alarm clock now, how similar is your alarm clock to the Sony clock in this study? 1 = Not at all 6 = Exactly the same	2.9	1.8	2.3	1.9	3.4	2.5
Have you ever used this alarm clock before? (% yes)	8.3%	–	16.7%	–	8.3%	–
How recently have you used an alarm clock before this study? 1 = Earlier today 7 = More than a year ago	1.9	1.7	2.3	1.7	3.6	2.2
How frequently do you use an alarm clock? 1 = Every day 7 = Only used once or twice	3.3	1.5	3.5	1.5	3.6	0.7

Source: O' Brien, M. A. (2010). *Understanding human–technology interactions: The role of prior experience and age.* PhD dissertation, Georgia Institute of Technology, Atlanta. With permission.

[a] A participant with missing data later referenced their own alarm clock as the source of knowledge for a specific clock control, so it seems reasonable to assume that the participant had actually used an alarm clock before.

and to older people because the large, easy-to-read display could moderate the effects of age-related declines in vision (e.g., Fisk, Rogers, Charness, Czaja, & Sharit, 2009). Note that the controls are the switches, buttons, and dials indicative of the electromechanical generation of technologies that predominated in the 1960s (Docampo Rama, 2001).

The primary of the clock in Figure 5.1 view shows the display with clear contrast for typical displays such as time, AM/PM, radio dial, and type of alarm set. Insets provide detailed views of the relevant controls for the tasks reported here. The top-left inset labeled "ALARM TIME/CLOCK SET" sets the mode for the time to be set. The top-right inset controls labeled "TIME SET" are used for controlling the minute changes. Note that both of these controls have etched information such as the "FAST −," "+," and "FAST +" on the "TIME SET" controls. The bottom-left inset control was used for selecting "ALARM MODE." The bottom-right inset controls were used for radio power (RADIO ON/RADIO OFF). The side inset controls were used for tuning and volume controls. Note that labels for both controls are only raised in the same silver color as the panel.

Participants were video-recorded in a comfortable laboratory setting. They were first oriented to everyday technologies by watching a PowerPoint slide show of representative technologies such as vending machines while the experimenter highlighted the fact that users typically completed tasks on these devices without training. To provide some familiarity with the technology and reduce stress about their performance, participants were asked to pick up the clock radio and point out the features they noticed as they briefly examined it. They were then asked to think aloud while completing three tasks using the clock:

1. Setting the clock to the current time (*set time*)
2. Changing the radio to a specified frequency, listening to the music, and adjusting the volume before turning the radio off (*radio volume*)
3. Setting the alarm to wake to radio at a specified time but leaving the clock displaying the correct time (*set alarm*)

After completing the tasks, they then completed a device usage questionnaire to document potentially relevant prior knowledge that may have been used during their interactions. On this questionnaire, participants evaluated their familiarity with the appearance, location, and operation of each of the six clock radio features using a scale adapted from Blackler, Mahar, and Popovic (2010). They also reported technologies from which they knew each feature, as per Blackler, Mahar, and Popovic, and the recency and frequency of access of each of these technologies. Lastly, they completed the NASA Raw Task Load Index (NASA-RTLX) cognitive workload questionnaire that included user perceptions of their success and satisfaction with their performance.

5.3.2 Results

Consistent with the diary study and with Blackler, Mahar, & Popovic (2010) and Reddy et al. (2010) but contrasting with Langdon et al. (2007), general technology experience did not eliminate age-related differences in performance (O'Brien, 2010). As will be described in this section, age differences were due both to the differential use of information on the device and to typical age-related cognitive and perceptual declines. In addition, low technology experience did not prevent several older people in this group from successfully completing all tasks. Surprisingly, there were few differences between the older groups classified by recent technology experience. In fact, many high-tech older people seemed to use the same reflective approach to completing tasks as low-tech older people when they noticed mistakes in their interaction. Both of these major findings suggest opportunities to improve technology usage through the appropriate design of device information by following guidance about improving usage for older people (e.g., Fisk et al., 2009). This section first reports sources of prior knowledge with several implications for design. Then, participants' subjective perceptions of their performance for each group are reported, followed by objective results from each task. Within these results, differences between performance that indicate intuitive and reflective cognition are described.

5.3.2.1   Prior Knowledge Used

Responses on the device usage questionnaire indicated technologies that provided prior knowledge to help participants use each control. Younger participants generally reported radio- or music-playing devices for controls used for the *radio volume* task and different clocks for controls used for the *set time* and *set alarm* tasks. Older participants reported a wider variety of devices for all controls, though the majority of devices were from the same domains as those reported by younger participants. The narrowness of these listings, even for very common controls such as a power button, was somewhat surprising, but the reports support Kirsh's (2013) contention that affordances are not just sensorimotor but part of an enactive landscape in which task goals shape our attention and thus prime the possibilities for actions and tasks enabled by controls that we see and touch. This finding extends the description in Chapter 2 of affordances as embodied knowledge (Blackler et al., 2018).

The new framework for intuitive interaction in Chapter 2 (Blackler et al., 2018) indicates that features that closely match a user's experience in the same context and for the same task will have a lower transfer distance and are thus more likely to be intuitively used, so transfer distances for each dimension of the relevant controls were assessed. The three dimensions of interest were *appearance*, *location*, and *operation*. Participants were asked to rate how much each control looked as they expected (appearance), was located where they expected (location), and worked as they expected (operation) on a scale of 1 to 6, where 6 was a perfect fit.

Table 5.2 presents dimension ratings for each participant group and success for the critical task in which the control was used. Different ratings for the different dimensions among successful participants suggest sensitivity to these dimensions during their interaction. For example, the fit of the *radio volume* controls, presented in the top rows of the table, showed strong overall similarity to controls on previously used devices, though the location of the radio power button was only moderately where participants expected. Unusual control locations slow visual searches because spatial location is the most salient cue for searches, and multiple unusual locations on a device may decrease users' confidence that the device can be used intuitively. Good fit of the control's appearance (including labeling) and operation (including expected feedback) can, however, increase users' feeling of familiarity, such that they can use fill-in strategies to guide correct control activation (Bastick, 2003; Beck, Peterson, & Vomela, 2006; Tsosyks & Gilbert, 2004).

TABLE 5.2
Ratings for Fit of Dimensions of Clock Radio Controls

Control	Dimension on Device Questionnaire	Younger People		High-Tech Older People		Low-Tech Older People	
		Not Successful[a]	Successful[a]	Not Successful[a]	Successful[a]	Not Successful[a]	Successful[a]
		(n=0)	(n=12)	(n=0)	(n=12)	(n=0)	(n=12)
Radio on	Appearance		4.50		4.91		5.08
	Location		3.83		4.73		4.42
	Operation		5.83		5.55		5.58
	% Similar function to another device		91.67%		90.91%		83.33%
Volume	Appearance		5.67		5.83		5.50
	Location		5.33		5.58		5.42
	Operation		5.92		5.92		5.58
	% Similar function to another device		100.00%		91.67%		91.67%
Tuner	Appearance		5.92		5.75		5.08
	Location		5.58		5.67		5.08
	Operation		5.92		5.92		5.42
	% Similar function to another device		100.00%		91.67%		91.67%

(Continued)

TABLE 5.2 (CONTINUED)
Ratings for Fit of Dimensions of Clock Radio Controls

Control	Dimension on Device Questionnaire	Younger People		High-Tech Older People		Low-Tech Older People	
		Not Successful[b]	Successful[b]	Not Successful[b]	Successful[b]	Not Successful[b]	Successful[b]
		(n=4)	(n=8)	(n=6)	(n=6)	(n=7)	(n=5)
Alarm time/clock set	Appearance	4.75	4.75	4.14	5.20	4.14	3.40
	Location	4.75	4.88	4.14	5.00	4.29	3.60
	Operation	5.25	5.25	4.14	5.60	4.86	3.80
	% Similar function to another device	75.00%	75.00%	100.00%	100.00%	71.43%	100.00%
Time set	Appearance	5.25	4.25	3.43	5.20	4.43	3.40
	Location	5.50	5.13	4.14	5.40	4.71	3.80
	Operation	4.25	4.63	4.14	5.80	4.43	3.40
	% Similar function to another device	100.00%	87.50%	100.00%	100.00%	85.71%	80.00%
Alarm mode	Appearance	5.00	4.50	4.29	5.60	4.00	4.00
	Location	5.00	3.75	4.29	5.60	4.25	4.40
	Operation	5.00	5.25	4.86	5.80	4.00	3.80
	% Similar function to another device	50.00%	62.50%	85.71%	80.00%	50.00%	60.00%

Source: O' Brien, M. A. (2010). *Understanding human–technology interactions: The role of prior experience and age.* PhD dissertation, Georgia Institute of Technology, Atlanta. With permission.

Note: Rating on a 1–6 scale, where 1 is no fit and 6 is a perfect fit.

[a] Success on the Radio Volume task.
[b] Success on the *Set Alarm* task.

These findings are generally consistent with Blackler, Popovic, and Mahar's (2010) examination of the benefit of location, appearance, or a combination of the two dimensions on intuitive usage. The study's finding was that the combination led to the highest intuitive usage, but appearance was more important for correct and intuitive uses as well as faster times than location, which may be because the presence of familiarity in two dimensions facilitated implicit access to the correct prior knowledge. If expectations for operation are also met, users are likely to perceive that they are progressing toward the goal, which increases feelings of familiarity and confidence in proceeding with a trial-and-error strategy. As detailed earlier in the description of metacognitive knowledge in the diary study, users frequently perceive this quick trial-and-error strategy as intuitive because they are unaware of the prior knowledge and reasoning that led to their choice of actions.

5.3.2.2 Subjective Perceptions of Performance

Subjective measures of user performance were particularly important to examine for the clock radio because the technology and tasks were expected to be familiar for all participants. Overall, mean cognitive workload ranged from 2.85 for younger people to 4.67 for low-tech older people on a scale of 1 (low) to 10 (high), with somewhat higher mental and perceptual activity as the highest related workload component for all groups. Importantly, no group reported perceiving strong time pressure, insecurity, or discouragement while completing these clock radio tasks (O'Brien, 2010). On the other end of the scale, participant perceptions of their success and satisfaction with their performance on the clock radio ranged from a mean of 6.83 for low-tech older people to 8.67 for younger people, with somewhat higher success than satisfaction ratings. No age or experience differences were identified between groups for any of these ratings based on Mann–Whitney U tests (all $p > .05$). These results suggest that participants were generally satisfied with their performance, perhaps similar to their real performance completing these tasks on a different clock radio. Thus, findings about the successes, problems, and strategies for this device are likely to represent the typical user approach for completing tasks on a new exemplar of a familiar technology.

5.3.2.3 Objective Results for High Familiarity Tasks

Similar performance on the *radio volume* task across all groups was revealed both through Mann–Whitney U tests of task success and t-tests of task completion time (all $p > .05$) (O'Brien, 2010). The task was successfully completed by all participants, and two or three in each group performed optimally without error. Thus, most participants made errors from which either prior knowledge or effective feedback guided quick recovery. Similar errors with the same controls were made by some participants in each group, such as incorrectly using the alarm mode button (bottom left in Figure 5.1), which suggests interference from devices that had been used by participants across ages and general technology experience or a level of stress and/or inability to think of any other solution to try. Overall, though, Table 5.2 shows similarly high fit ratings for each dimension of the critical *power, volume,* and *tuner* controls with participant expectations among all groups. These similarities indicate the effect of substantial task knowledge to guide users in a trial-and-error interaction even on a new implementation of the task. These ratings thus suggest that users perceive their interaction as intuitive because actions are guided by familiarity and affordances rather than conscious awareness of the next step and expected feedback. The placement of controls and displays in close proximity to each other on the right of the clock ("VOL" and "TUNE") or on the bottom right ("RADIO ON" and "RADIO OFF") of the clock front, as shown in Figure 5.1, also allowed even older participants with lower working memory and attentional control to notice error feedback and correctly interpret it to try a different option in an intuitive manner.

5.3.2.4 Age Differences in Less Familiar Tasks

Multiple age differences in the *set time* and *set alarm* tasks, which required the use of controls on the top of the clock (top-right and left insets in Figure 5.1) and display feedback on the black face of the clock (main image in Figure 5.1), were identified. Mann–Whitney U tests identified no age differences in task success, but t-tests revealed that younger participants were significantly faster than high-tech older participants on the *set alarm* task ($t[22] = -3.27$, $p < .05$) (O'Brien, 2010). In general, results showed large inter-individual differences among the high-tech older participants. Some of these participants performed optimally and nearly as quickly as the younger participants, but some participants experienced multiple problems with multiple controls from which recovery was not always successful. Dimension expectations for high-tech older participants shown in Table 5.2 also suggest better fit with expectations for successful compared with unsuccessful participants. Thus, lower transfer distance may be an effective measure to indicate whether features could be intuitively used, but all three dimensions of the feature must be measured. Although Blackler, Popovic, and Mahar's (2010) research reports that using familiar features for two of the dimensions increases intuitive usage, designers should modify their design if the transfer distance for the third dimension is significantly higher than the others, because the anomaly may elicit focused attention and reflective usage.

Equivalent levels of performance were demonstrated for the *set time* task with no age differences in task time, though high-tech older participants performed more actions overall ($t[22] = -1.64$, $p < .05$). High-tech older participants made some of the same errors as younger participants, but more older participants made these errors multiple times, misread more feedback from different places, and operated several unrelated controls. Yet, equivalent task times between groups suggests that even high-tech older participants used a trial-and-error strategy with quick and effortless feedback assessment and next action selection to complete this task. As with the younger participants, they would also likely perceive this interaction as intuitive because actions were guided by high familiarity with the controls rather than conscious awareness of the next step and expected feedback.

Differences within the *set alarm* task, however, suggest that these unsuccessful older participants did not learn basic clock operations from their experience. Overall, high-tech older participants were slower ($t[22] = -3.27$, $p < .05$), made more errors ($t[22] = -2.88$, $p < .01$), and performed more actions ($t[22] = -3.75$, $p < .01$) on this task (O'Brien, 2010). This result is similar to the results of Lawry, Popovic, and Blackler (2011), who found that older people using cameras and alarm clocks in a lab setting were also less able than younger people to learn from their experience as they progressed though the set tasks. Completing this task also required using the alarm mode controls on the black face of the clock (bottom-left inset of Figure 5.1), but several participants in both high-tech groups failed to use this control correctly and thus only partially completed the task. Younger people also operated other unrelated controls and misread feedback, such that errors were nearly double those for the *set time* task, but they generally demonstrated learning the basic operations of this clock radio.

Several problems were observed among high-tech older people but not among younger people. For example, several older participants repeated problems from the first task, including perseveration with the same incorrect interaction style on the time set controls (top-right inset in Figure 5.1). This repetition of an incorrect procedure by older people in an experimental setting was also observed by Blackler (2006), Harada and Suto (2008), and Kang and Yoon (2008), and researchers have suggested it may be due to episodic memory declines that reduce the user's recall of prior actions. Overall, older people also used more unrelated controls and misread more feedback, suggesting that age-related cognitive and perceptual declines may have hindered their ability to recover. For example, researchers note that older people have more difficulty coordinating button presses and monitoring feedback when they are on different surfaces (e.g., Watson, Lambert, Cooper, Boyle, & Strayer, 2013). Thus, design changes sensitive to aging users could have improved performance on this task.

One critical age difference in these interactions was high-tech older people's higher use of "looks" for both the *radio volume* and *set alarm* tasks (O'Brien, 2010). Looks were counted by researchers during analysis of video recordings whenever participants examined or described information on the device *except* immediately before activating a control. Looks were typically used by all participants to verify control settings or to read a label presented in low contrast, such as the "FAST −," "+," and "FAST +" time set controls in the top-right inset of the clock shown in Figure 5.1. As an interaction strategy, however, older people's looks suggested a reflective process whereby they identified a task step, examined a control label that seemed to match the step description in their mind, and identified the expected response to activating the control before actually making the selection. This process includes the controlled, effortful, deductive, and self-aware characteristics of reflective cognition (Kahneman & Frederick, 2002). Thus, the observed looks provide evidence that a high technology experience level is insufficient to elicit intuitive usage for any particular technology by every user, even if some users do use it intuitively.

This reflective strategy by which users focus step by step on goal identification, label examination, and expected feedback for each control before activation was identified by Polson and Lewis (1990) from their early review of menu-based technology interactions to describe how users determine what to do next on new technologies. They called this approach *label-following*, but this description seems to fit the high use of looks by older participants in our study as well. Of course, determining what to do next is critical when interacting with any technology, but this label-following based on looking carefully before acting seems to have been developed by users to guard against errors they could not resolve. An alternative approach that mitigates the need for label-following uses device feedforward as an effective mechanism to naturally guide users to the next step with low effort and precon-scious anticipation during control selection (e.g., Djajadiningrat, Wensveen, Frens, & Overbeeke, 2004; Larkin, McDermott, Simon & Simon, 1980). Feedforward in these descriptions taps implicit processing by leveraging feelings of semantic and articulatory directness that indicate a close match of the environment with the con-text for well-learned activities. Thus, effective feedforward operates naturally with sensorimotor skills that control our embodied interaction with the physical world, as described in Chapter 2 (Blackler et al., 2018), including coordinated timing and dynamics with other concurrent actions that mimic the rhythm and flow of natural actions and responses.

To reduce older people's preference for avoiding errors with the slower reflec-tive approach, designers could create alternative feedforward mechanisms for older people based on sensorimotor skills that can build user confidence and guide more fluid activity. For example, a video game designed to provide therapy for frail older people used movement-based guidance and biofeedback when the user was predicted to fall by the system sensors. Within one or two cycles of practice, the users implic-itly learned which movements predicted falls, and thus they anticipated their own balance loss and unconsciously made the adjustments themselves (Szturm, Betker, Moussavi, Desai, & Goodman, 2011). This result supports the opportunity for elic-iting intuitive interaction through effective feedforward design, even among older people in a task associated with high error consequences.

5.3.2.5   Experience Differences in Less Familiar Tasks

Several experience differences were identified between low- and high-tech older peo-ple in the *set time* and *set alarm* tasks. Specifically, Mann–Whitney U tests revealed that high-tech older people were significantly more likely to successfully complete the *set alarm* tasks ($U = 25$, $p < .005$, $r = .65$), but low technology experience did not prevent several older people in the low-tech group from successfully completing all the tasks (O'Brien, 2010). High-tech older people were significantly faster than low-tech older people only in the *set time* task, as revealed by t-tests ($t[22] = 2.60$, $p < .05$). This slower time represents low-tech older participant comments such as "I don't just push buttons unless I know what they are for." Without manuals or other sources of information, participants therefore resorted to the same reflective strategy reported by low-tech older participants in the diary study. Findings of the consistent use of reflective strategies from both self-report and observational studies suggest a deliber-ate trade-off of accuracy/error avoidance for speed among this population.

Comparing control dimension ratings for unsuccessful with successful low-tech older people in Table 5.2, successful participants completed the task even though fit was only moderate, with small differences among the dimensions. Relevant control dimension ratings for unsuccessful low-tech participants were higher, perhaps indi-cating that they did not notice relevant differences that they should have noticed in order to recover from errors. As described previously, successful high-tech older people may also switch from their default trial-and-error approach to a reflective strategy if they recognize that they are not making progress without slowing down and considering their next action more carefully. The slightly higher success rate and lower number of errors for high-tech older people also suggests a better ability to interpret feedback and a higher knowledge of possible technology strategies to use for recovery than low-tech older people. Prior research (Hammond et al., 1987) suggested that individuals switch between intuitive and reflective processing during more complex tasks, and high-tech older people may have had experience with technology that familiarized them with how to implicitly determine which cognitive processing mode should be used.

5.4 DISCUSSION AND DESIGN IMPLICATIONS

The purpose of presenting this research about everyday technology interactions to inform research on intuitive interactions was to systematically explore the use of prior knowledge and familiarity among younger and older people. Our research showed that general technology experience predicts users' breadth of experience with computer and mobile device features and functions. This technical knowledge, along with knowledge of common problem-solving strategies for technologies, helped older people with high technology experience to resolve more problems than same-aged people with low technology experience.

Perhaps as important as technical knowledge, we also found that the metacognitive knowledge for each cohort predicted technology attitudes and behaviors. For example, high-tech older participants demonstrated interest in learning new technologies and features even during their everyday lives in the diary study (O'Brien, 2010). As discussed in Section 5.3.2, these differences seemed to translate in the laboratory study into different first-use strategies for the familiar clock radio, such that high-tech older participants began interactions using an intuitive trial-and-error approach like younger people, but low-tech older participants started with the reflective approach that was also reported in the diary study for diverse everyday technology interactions. When high-tech older participants experienced problems, however, they also seemed to switch from the intuitive approach to the reflective approach. Unlike younger participants, though, high-tech older participants were more likely to blame their own abilities or settle for lower performance levels with a clock time that was "close enough" but not correct. Compared with low-tech older participants with preferences for limited technology use in general, though, the resulting reluctance of the high-tech older person to use technologies with which they have problems is focused on the specific function they could not complete. As reported in the diary study, high-tech older people were more likely than younger people to immediately retrieve an instructional manual for completing low-frequency tasks so that time was not wasted trying to recall specific steps or recover from errors.

Both studies reported here also identified the importance of understanding functional knowledge among target populations. Familiarity with task goal, critical steps, and relevant functions to use in the *radio volume* task for the laboratory study allowed even low-tech older participants to perform successfully. The appropriate control for that step could be quickly detected through automatic visual search processes, so participants did not stop to think about the next step as they would during reflective processing. Indeed, prior research has suggested that high confidence and knowledge of the task goal but low confidence and knowledge in the task method leads to intuitive processing (Hammond et al., 1987). This intuitive processing may occur because the task goals implicitly prime relevant functions and task knowledge such as the first step in the task sequence (Kirsh, 2013).

Note, however, that the effective detection of expected controls and feedback also depends on the control being in a familiar location; otherwise, visual search processes will be interrupted (e.g., Beck et al., 2006). If the controls also operate as expected, especially providing familiar tactile and audible feedback that can be implicitly registered by sensorimotor systems, users can confirm the task is proceeding toward the goal with little attention. Thus, familiar appearance, locations, and operations from prior devices used in a new technology's functional domain may indicate relevant prior knowledge that designers should consider, as indicated in the Enhanced Framework for Intuitive Interaction presented in Chapter 2 (Blackler et al., 2018). As our research has shown, however, reflective label-following can be used to complete the task at a slower but more accurate rate. The broader preference for older users to sacrifice speed for accuracy (e.g., Fisk et al., 2009) may thus be appropriate for them to successfully complete tasks, given rapidly changing technologies as well as the cognitive and physical declines they may be facing.

The specific effects of age differences alone were also identified in these studies. First, they correspond to the types of knowledge just highlighted, as they indicate the typical feature knowledge of the relevant technology generation (Docampo Rama, 2001), technology self-efficacy (e.g., Lagan, Oliver, Ainsworth, & Edwards, 2011), and cultural knowledge, such as options for alternative ways to complete a task (e.g., O'Brien & Rogers, 2013). Second, typical perceptual and cognitive declines that accompany aging were indicated by the errors made in both studies. Third, good design practices seem to mitigate these declines and facilitate older users in completing the task using extant functional knowledge. For example, completion of the *radio volume* task was facilitated by clear progress indicators on the display that were close to the controls, allowing users to immediately recognize when a control did not provide the expected function and they should try another. Thus, older participants did not experience the feeling of "lostness" that increases older people's anxiety and reluctance to use technology (Barnard et al., 2013).

Findings such as these support an increased dependence on adherence to effective design principles because older people rely on this visual guidance and feedback even more than younger people (e.g., Spieler, Mayr, & Lagrone, 2006). For example, an alarm clock design strictly following the proximity principle (e.g., Zheng, 2013), which suggests that feedforward and feedback about related functions should be close together, would likely have increased older users' ability to correct problems and complete the *set alarm* task. Designers may not realize that older users are more likely to have interference issues with multiple instances of *set time* controls that have to be individually confirmed or denied to allow participants to proceed with the task. When the older user misses feedback indicating an incorrect hypothesis because of their smaller functional field of view (e.g., Ball, Beard, Roenker, Miller, & Griggs, 1988), they typically proceed to the next step incorrectly. Given the incorrect predecessor step, they are likely to only stop later when they reach an impasse without knowledge of which step to return to and correct. Thus, even though users could technically recover, their disorientation and lower self-efficacy may make this recovery quite unlikely.

Beyond careful attention to design best practices, designers may be able to create technology coaches (e.g., Blanson Henkemans, Rogers, Fisk, Neerincx, Lindenberg, & van der Mast, 2008) that support first-time use in order to increase the adoption of contemporary technologies, which will be particularly essential in the health-care realm that is vitally important to older people's independence. To create intuitive technologies that provide innovative and magical experiences for older people, however, new research may be needed to specify best practices for providing the metacognitive support for orientation and control currently lacking. As Howell (1997) noted over 20 years ago, "Often the aging process simply exacerbates subop-timal design features that, to a lesser extent, affect everyone's performance" (p. 7). Moreover, younger participants also made some of the same errors as high-tech older participants in the reported studies and could thus also benefit from more robust environmental support.

5.5 CONCLUSION

Findings reported in this chapter suggest that older people may actually want to use new technologies, but their experience suggests to them, at least implicitly, that their prior knowledge and familiarity with other technologies can hurt their effectiveness. Thus, their preference for having instruction or another person initially orient them to a new technology (Anderson & Perrin, 2017) seems perfectly adaptive.

The rapid pace of changing technologies makes it unlikely that training programs can keep up with these changes, so designers must consider how to enable first-time use even for new exemplars of existing technologies. Successful first uses need not be error free, but users must feel safe in making errors so that they will not get lost or create problems for other people. Feedforward that taps extant sensorimotor skills may deter the preference for instruction-led first use and build user knowledge and confidence for exploring new features of the technology. Feedback that enables rapid problem detection and recovery can help users feel confident performing basic operations and want to explore additional features in subsequent uses. Importantly, feelings of familiarity and directness that accompany progress toward the goal can maintain the metacognitive decision to interact intuitively with a specific device in a specific context, even in spite of errors. Thus, the fundamental lesson of safe explo-ration from a guided first use may thus be the most important factor for increasing intuitive interactions among a broader population.

ACKNOWLEDGMENTS

This chapter is based on the author's doctoral dissertation (O'Brien, 2010). Portions of these data were presented at the Southern Gerontological Society Annual Meeting (O'Brien, Rogers, & Fisk, 2009a), the 62nd Annual Meeting of the Gerontological Society of America (O'Brien, Rogers, & Fisk, 2009b), and the 2010 Cognitive Aging Conference (O'Brien, Burnett, Rogers & Fisk, 2010). This research was supported in part by a grant from the National Institutes of Health (National Institute on Aging) (Grant P01AG17211) under the auspices of the Center for Research and Education on Aging and Technology Enhancement (CREATE; www.create-center.org) and in part by a Ruth L. Kirschstein National Research Service Award (NRSA) Institutional Research Training Grant from the National Institutes of Health (National Institute on Aging) (Grant T32AG000175). The research could not have been done without the guidance of Wendy A. Rogers and Arthur D. Fisk.

REFERENCES

Anderson, M., & Perrin, A. (2017). *Tech adoption climbs among older adults.* Pew Research Center, May 17. Retrieved from: http://www.pewinternet.org/2017/05/17/tech-adoption-climbs-among-older-adults, accessed May 23, 2017.

Ball, K. K., Beard, B. L., Roenker, D. L., Miller, R. L., & Griggs, D. S. (1988). Age and visual search: Expanding the useful field of view. *Journal of the Optical Society of America A, 5*(12), 2210–2219.

Barnard, Y., Bradley, M. D., Hodgson, F., & Lloyd, A. D. (2013). Learning to use new technologies by older adults: Perceived difficulties, experimentation behaviour and usability. *Computers in Human Behavior, 29*(4), 1715–1724.

Bastick, T. (2003). *Intuition: Evaluating the Construct and Its Impact on Creative Thinking.* Kingston, Jamaica: Stoneman and Lang.

Beck, M. R., Peterson, M. S., & Vomela, M. (2006). Memory for where, but not what, is used during visual search. *Journal of Experimental Psychology: Human Perception and Performance, 32,* 235–250.

Blackler, A. (2006). *Intuitive interaction with complex artefacts.* PhD dissertation, Queensland University of Technology, Brisbane, Australia.

Blackler, A. (2018). Intuitive interaction: An overview. In A. Blackler (Ed.), *Intuitive Interaction: Research and Application* (pp. 3–18). Boca Raton, FL: CRC Press.

Blackler, A., Desai, S., McEwan, M., Popovic, V., & Diefenbach, S. (2018). Perspectives on the nature of intuitive interaction. In A. Blackler (Ed.), *Intuitive Interaction: Research and Application* (pp. 19–40). Boca Raton, FL: CRC Press.

Blackler, A., Mahar, D., & Popovic, V. (2010). Older adults, interface experience, and cognitive decline. Paper presented at OZCHI 2010, "Design—Interaction—Participation," 22nd Annual Conference on the Australian Computer–Human Interaction Special Interest Group, Brisbane, Australia.

Blackler, A., Popovic, V., & Mahar, D. (2010). Investigating users' interaction with complex artefacts. *Applied Ergonomics, 41*(1), 72–92.

Blanson Henkemans, O. A., Rogers, W. A., Fisk, A. D., Neerincx, M. A., Lindenberg, J., & van der Mast, C. A. P. G. (2008). Usability of an adaptive computer assistant that improves self-care and health literacy of older adults. *Methods of Information in Medicine, 47,* 82–88.

Byrne, M. D., & Davis, E. M. (2006). Task structure and postcompletion error in the execution of a routine procedure. *Human Factors: The Journal of the Human Factors and Ergonomics Society, 48*(4), 627–638.

Chen, Z., & Mittel, A. V. (2006). Generalization and transfer of problem solving strategies. In *Focus on Educational Psychology* (pp. 217–234). Hauppauge, NY: Nova Science.

Djajadiningrat, T., Wensveen, S., Frens, J., & Overbeeke, C. (2004). Tangible products: Redressing the balance between appearance and action. *Personal and Ubiquitous Computing, 8,* 294–309.

Docampo Rama, M. (2001). *Technology Generations: Handling Complex User Interfaces.* PhD dissertation, Technical University of Eindhoven, the Netherlands.

Fisk, A. D., Rogers, W. A., Charness, N., Czaja, S. J., & Sharit, J. (2009). *Designing for Older Adults: Principles and Creative Human Factors Approaches* (2nd ed.). Boca Raton, FL: CRC Press.

Flavell, J. H. (1979). Metacognitive and cognitive monitoring: A new area of cognitive developmental inquiry. *American Psychologist, 34*, 906–911.

Freudenthal, A., & Mook, H. J. (2003). The evaluation of an innovative intelligent thermostat interface: Universal usability and age differences. *Cognition, Technology, and Work, 5*, 55–66.

Hammond, K. R., Hamm, R. M., Grassia, J., & Pearson, T. (1987). Direct comparison of the efficacy of intuitive and analytical cognition in expert judgment. *IEEE Transactions on Systems, Man, and Cybernetics, 17*(5), 753–770.

Harada, E. T., & Suto, S. (2008). Error repetitions and cognitive control: Why does it happen outside the psychology lab? Paper presented at the 2008 Cognitive Aging Conference, Atlanta, GA.

Howell, W. (1997). Forward, perspectives, and prospectives. In A. D. Fisk & W. A. Rogers (Eds.), *Handbook of Human Factors and the Older Adult* (pp. 1–6). San Diego, CA: Academic Press.

Kahneman, D., & Frederick, S. (2002). Representativeness revisited: Attributed substitution in intuitive judgment. In T. Gilovich, D. Griffin, & D. Kahneman (Eds.), *Heuristics and Biases: The Psychology of Intuitive Judgment* (pp. 49–81). Cambridge, UK: Cambridge University Press.

Kang, N. W., & Yoon, W. C. (2008). Age- and experience-related user behavior differences in the use of complicated electronic devices. *International Journal of Human–Computer Studies, 66*, 425–437.

Kirsh, D. (2013). Embodied cognition and the magical future of interaction design. *ACM Transactions on Computer–Human Interaction (TOCHI), 20*(1), 3.

Laganà, L., Oliver, T., Ainsworth, A., & Edwards, M. (2011). Enhancing computer self-efficacy and attitudes in multi-ethnic older adults: A randomised controlled study. *Ageing and Society, 31*(06), 911–933.

Langdon, P., Lewis, T., & Clarkson, J. (2007). The effects of prior experience on the use of consumer products. *Universal Access in the Information Society, 6*, 179–191.

Larkin, J., McDermott, J., Simon, D. P., & Simon, H. A. (1980). Expert and novice performance in solving physics problems. *Science, 208*, 1335–1342.

Lawry, S. (2012). *Identifying familiarity to facilitate intuitive interaction for older adults.* PhD dissertation, Queensland University of Technology, Brisbane, Australia.

Lawry, S., Popovic, V., & Blackler, A. (2011). Diversity in product familiarity across younger and older adults. Paper presented at IASDR 2011, Fourth World Conference on Design Research, Delft, the Netherlands.

O'Brien, M. A. (2010). *Understanding human–technology interactions: The role of prior experience and age.* PhD dissertation, Georgia Institute of Technology, Atlanta.

O'Brien, M. A., Burnett, J. S., Rogers, W. A., & Fisk, A. D. (2010). Understanding age-related problems in use of everyday technologies. Presented at the 2009 Cognitive Aging Conference, April 15–18, St Petersburg, FL.

O'Brien, M. A., Rogers, W. A., & Fisk, A. D. (2009a). Discerning prior knowledge use by older adults in human technology interactions. Presented at the 2009 Southern Gerontological Society Annual Meeting, April 16–19, St Petersburg, FL.

O'Brien, M. A., Rogers, W. A., & Fisk, A. D. (2009b). The role of prior experience in everyday technology interactions. Presented at the 62nd Annual Scientific Meeting of the Gerontological Society of America, November 18–22, Atlanta, GA.

O'Brien, M. A., Rogers, W. A., and Fisk, A. D. (2010). Developing an organizational model for intuitive design. Technical Report HFA-TR-100. Georgia Institute of Technology, School of Psychology, Human Factors and Aging Laboratory, Atlanta. Retrieved from: http://smartech.gatech.edu/bitstream/handle/1853/40563/HFA-TR-1001IntuitiveDesignConceptualOverview.pdf?sequence=1 (accessed October 26, 2011).

O'Brien, M. A., & Rogers, W. A. (2013). Design for aging: Enhancing everyday technology use. In R. Z. Zheng, R. D. Hill, & M. K. Gardner (Eds.), *Engaging Older Adults with Modern Technology: Internet Use and Information Access Needs* (pp. 105–123). Hershey, PA: IGI.

O'Brien, M. A., Rogers, W. A., & Fisk, A. D. (2012). Understanding age and technology experience differences in use of prior knowledge for everyday technology interactions. *ACM Transactions on Accessible Computing, 4*(2), 2.

O'Brien, M. A., Weger, K., DeFour, M. E., & Reeves, S. M. (2011). Examining the role of age and experience on use of knowledge in the world for everyday technology interactions. In *Proceedings of the Human Factors and Ergonomics Society 55th Annual Meeting* (pp. 177–181). Santa Monica, CA: Human Factors and Ergonomics Society.

Olson, K. E., O'Brien, M. A., Rogers, W. A., & Charness, N. (2011). Diffusion of technology for younger and older adults. *Ageing International, 36*(1), 123–145.

Payne, S. J. (2003). User's mental models: The very ideas. In J. M. Carroll (Ed.), *HCI Models, Theories, and Frameworks* (pp. 135–156). Amsterdam, the Netherlands: Morgan Kaufmann.

Polson, P. G., & Lewis, C. H. (1990). Theory-based design for easily learned interfaces. *Human-Computer Interaction, 5*, 191–220.

Reddy, R. G., Blackler, A., Mahar, D., & Popovic, V. (2010). The effect of cognitive ageing on use of complex interfaces. Paper presented at OZCHI 2010, "Design—Interaction—Participation," 22nd Annual Conference on the Australian Computer–Human Interaction Special Interest Group, Brisbane, Australia.

Rogers, W. A., O'Brien, M. A., & Fisk, A. D. (2013). Cognitive engineering to support successful aging. In J. D. Lee, A. Kirlik, & M. J. Dainoff (Eds.), *The Oxford Handbook of Cognitive Engineering* (pp. 286–301). New York: Oxford University Press.

Singley, M. K., & Anderson, J. R. (1987). A keystroke analysis of learning and transfer in text editing. *Human–Computer Interaction, 3*, 223.

Sloman, S. A. (1996). The empirical case for two systems of reasoning. *Psychological Bulletin, 119*, 3–22.

Spieler, D. H., Mayr, U., & LaGrone, S. (2006). Outsourcing cognitive control to the environment: Adult age differences in the use of task cues. *Psychonomic Bulletin & Review, 13*, 787–793.

Still, J., & Still, M. (2018). Cognitively describing intuitive interactions. In A. Blackler (Ed.), *Intuitive Interaction: Research and Application* (pp. 41–62). Boca Raton, FL: CRC Press.

Szturm, T., Betker, A. L., Moussavi, Z., Desai, A., & Goodman, V. (2011). Effects of an interactive computer game exercise regimen on balance impairment in frail community-dwelling older adults: A randomized controlled trial. *Physical Therapy, 91*(10), 1449–1462.

Tsodyks, M., & Gilbert, C. (2004). Neural networks and perceptual learning. *Nature, 431*, 775–781.

Was, C. A., & Woltz, D. J. (2013). Implicit memory and aging: Adapting technology to utilize preserved memory functions. In R. Z. Zheng, R. D. Hill, & M. K. Gardner (Eds.), *Engaging Older Adults with Modern Technology: Internet Use and Information Access Needs* (pp. 1–19). Hershey, PA: IGI.

Watson, J. M., Lambert, A. E., Cooper, J. M., Boyle, I. V., & Strayer, D. L. (2013). On attentional control and the aging driver. In R. Z. Zheng, R. D. Hill, & M. K. Gardner (Eds.), *Engaging Older Adults with Modern Technology: Internet Use and Information Access Needs* (pp. 20–32). Hershey, PA: IGI.

Zheng, R. Z. (2013). Effective online learning for older people: A heuristic design approach. In R. Z. Zheng, R. D. Hill, & M. K. Gardner (Eds.), *Engaging Older Adults with Modern Technology: Internet Use and Information Access Needs* (pp. 142–159). Hershey, PA: IGI.

6 Adaptable Interface Framework for Intuitively Learnable Product Interfaces for People with Diverse Capabilities

Gudur Raghavendra Reddy,
Alethea Blackler, and Vesna Popovic

CONTENTS

6.1 INTRODUCTION

The intuitive use of an interface involves the non-conscious use of prior knowledge, and designing for intuitive use requires that an interface is designed to match the prior knowledge and competence of a user group. This process is effective when the target user group's prior experience is homogenous. However, it is challenging to implement this approach when designing for older people, mostly because of their diversity in terms of technology familiarity (TF) and cognitive abilities.

Much research has been done to address the problem of diversity in older user groups. However, most of these studies suggest two design paradigms: (1) exclusivity, a path that encourages designing separate products or interfaces for older users, and (2) inclusivity, which encourages designing a product that is universally usable.

113

A truly inclusive/universal design is ideal. However, a "one design fits all" approach is not practical simply because it is extremely difficult to anticipate the variability of a target group and the contexts of use at the time of its design (Akiki, Bandara, & Yu, 2014).

There is a significant body of research that expounds how adaptive interfaces could address the issues of user diversity and the context of product use (Akiki et al., 2014; Peissner & Edlin-White, 2013; Peissner, Häbe, Janssen, & Sellner, 2012; Shneiderman, 2003; Sloan, Atkinson, Machin, & Li, 2010; Stephanidis, 2001). Adaptive or adaptable interfaces originate from the belief that if one can anticipate a user's needs, it is possible to present an interface with only relevant options, thus simplifying or matching any given user's particular abilities or prior experience. However, at the same time, various studies also indicate that this could lead to usability problems (Akiki, Bandara, & Yu, 2013; Shneiderman & Plaisant, 2010; Stuerzlinger, Chapuis, Phillips, & Roussel, 2006).

Research on adaptive and adaptable interfaces spans decades, from one of the early discussions on the topic in the 1980s (Edmonds, 1981) to more recent work that talks about its importance to accessibility (Peissner & Edlin-White, 2013; Smirek, Zimmermann, & Beigl, 2016). During this period, many models and frameworks have been suggested. However, this has not yet resulted in a practical implementation of this paradigm (Lavie & Meyer, 2010; Peissner et al., 2012). Perhaps a reason could be because of the complexity of most of these suggestions. They often look at adaptation from the perspective of system-level implementation. Adaptation at that level requires a huge knowledge base of use contexts, patterns, widgets, or extremely complex algorithms (Leonidis, Antona, & Stephanidis, 2012; Peissner et al., 2012).

Here we present our research that addresses some of these challenges in the implementation of adaptable interface frameworks by taking a simpler, rule-based, designer-driven approach. This framework is developed to help in designing interfaces that are easier to learn and, over time, intuitive to use for older users and others with diverse capabilities. The framework was developed based on the outcomes of experiments that investigated the effects of interface interventions, age, cognitive functioning, and TF on the use of product interfaces by older people. The outcomes of these experiments and the relevant literature are briefly discussed to provide a background to the development of the Intuitively Learnable Adaptable Interface System (ILAIS; Figure 6.1).

6.2 BACKGROUND LITERATURE

6.2.1 Technology Familiarity, Intuitive Interaction, and Aging

One of the important reasons for older people finding contemporary interfaces difficult to use is their low prior experience or TF (Hurtienne, Horn, & Langdon, 2010; O'Brien, 2010). TF and competence with technologies related to the product are essential for the intuitive use of its interface (Blackler, 2008; Blackler, Popovic, & Mahar, 2010; Hurtienne, Weber, & Blessing, 2008; Lewis, Langdon, & Clarkson,

2008; Naumann et al., 2008). Studies have also shown that older adults (over 60) were less familiar with their own products than younger adults and that both older and middle-aged adults (over 45) were less familiar with products they did not own than those under 45. In addition, older people were faster at using an analogue product and younger people were faster with the digital equivalent (Lawry, Popovic, & Blackler, 2009, 2010, 2011). There are many reasons behind older people being deficient in prior experience and also more varied in their capabilities. For example, age-related cognitive decline (Blackler, Mahar, & Popovic, 2010; Langdon, Lewis, & Clarkson, 2007; Lim, 2009), low perceived self-efficacy (Bandura, Freeman, & Lightsey, 1999; Czaja & Lee, 2007) and cohort effects, whereby *technology generations* are formed based on population groups who were exposed to certain types of interfaces during their formative years (Docampo Rama, Ridder, & Bouma, 2001; Lim, 2009).

In addition, as people age, they tend to specialize in an area of their choice and their interests thus become more focused. Over a period of time, this brings about increased variability in older people compared with younger ones (Salthouse, 2010). Age-related cognitive decline slows down the acquisition of new knowledge (Bäckman, Small, & Wahlin, 2001). The awareness of this limitation possibly also compels older people to be more selective in determining what they should learn. Older people do not necessarily see a need to keep up with technology for the sake of doing so. However, where they do see a need, they embrace the technology without reservation (Czaja & Lee, 2007). These and other related factors result in a variability of knowledge and a reluctance or inability in older people to adopt contemporary technologies.

The basic tenet of designing an intuitive interface is that its functions and how they are represented should match target user groups' familiarity. In designing intuitive interfaces for older people, designers face two major problems. First, older people are a very heterogeneous group in terms of both prior experience and cognitive capabilities. Therefore, to discover the common knowledge of this target group before developing intuitive product interfaces could be a very complex and extremely resource-intensive exercise. Second, research has shown that cognitive abilities play a mediating role in use of TF (Blackler, Mahar, et al., 2010; Reddy, Blackler, Popovic, & Mahar, 2016). This finding suggests that it is also important to know, apart from their technology exposure and competence, how the target users make use of TF under different circumstances and situations. A person might have been exposed to a device but, for various reasons, may not be able to retrieve knowledge about it when required. As a result, it is difficult to develop product interfaces that are entirely intuitive to use for a group of people with diverse capabilities, such as older people.

6.2.2 Adaptive and Adaptable Interfaces

Interfaces that adapt while in use are generally termed *adaptive interfaces* (Downton, 1993). There are various mechanisms of adaptation: automatic system-based adaptation, adaptation by system–user cooperation, adaptation by an expert intermediary, and adaptation by the users themselves (Edmonds, 1981). In general, user interface adaptation can be categorized under the following three headings (Akiki et al., 2013):

1. *Adaptable*: interfaces manually adapted by concerned stakeholders
2. *Adaptive* (or *mixed initiative*): interfaces adapted automatically to the context of use based on rules
3. *Truly adaptive*: interfaces adapted to known and unknown contexts of use completely automatically

Truly automatic adaptation is generally what is referred to in the literature when the notion of adaptive system behavior is discussed. Conceptually, when compared with adaptable interfaces, truly adaptive interfaces hold greater promise for developing accessible or universal interface design (Stephanidis, 2001).

However, there are many adaptive and adaptable systems for generating interfaces, and most of them concentrate on adaptation based on user characteristics, environmental conditions, and devices/platforms (Akiki et al., 2014; Peissner et al., 2012). In addition, there are two dimensions of adaptation (Stephanidis, Paramythis, Karagiannidis, & Savidis, 1997):

1. *Time of adaptation*: whether the system adapts at the initiation of the interaction (adaptable) or during runtime (adaptive)
2. *Level of interaction*: at what point of interaction certain rules of adaptation should apply

In terms of system development, there are more commonalties than differences between truly adaptive and adaptable interface systems (Gullà, Ceccacci, Germani, & Cavalieri, 2015). An adaptive system, based on rules, reacts to a user's ability, the context of use, and the device it is used on by initiating an appropriate level of interface on the fly. An adaptable system, on the other hand, initiates (by the user or an intermediary) an appropriate level of interface at the beginning of an interaction. O'Brien (2018, Chapter 5) also suggests that designers may be able to create technology coaches that support first-time use, which could form part of the adaptation process.

Truly adaptive interfaces have often been criticized for usability issues (Peissner & Edlin-White, 2013; Shneiderman & Plaisant, 2010). Shneiderman and Plaisant (2010) note that there are two problems in the implementation of adaptive interfaces. First, it is not always possible to predict the user's current requirements based on their past performance. Second, an interface that changes based on the user's performance could be very disruptive, especially if they are already familiar with it. Earlier research that examined Microsoft's experience with adaptive menus in Office 2000 suggested that users preferred adaptable (user-controlled) interfaces to adaptive ones (Shneiderman & Plaisant, 2010).

The limitations of truly adaptive interfaces could be circumvented by minimizing drastic changes in the interface, such as adding options to a stable layout or keeping the overall information structure intact and only changing options that are similar in function. This is similar to an online newspaper layout with different news sections, where news content changes but it is always relevant to the section it is posted under

(Shneiderman & Plaisant, 2010). This suggestion is supported by research that compared three implementations of adaptations:

1. *Spatially stable*, where options are presented in a tool bar that did not move
2. *Moving interface*, where options are moved from inside a pop-up menu to a menu bar
3. *Visual pop-out*, where an option is promoted inside a pop-out menu

This research showed that users preferred the spatially stable adaptation (Gajos, Czerwinski, Tan, & Weld, 2006).

Findlater and McGrenere (2004), in a controlled lab study with 27 participants, compared efficiency on static, adaptable, and adaptive menus. They found significant support for adaptable/adaptive interfaces, when compared with truly adaptive interfaces, in a general population sample. In a related and more recent study, a multi-layered interface paradigm was tested on small-screen device with older participants (Leung, Findlater, McGrenere, Graf, & Yang, 2010). The outcome of this study suggests that older people benefit most from an adaptable interface. On the downside, it was also observed that when the interface was changed from reduced functionality to full functionality, it negatively affected performance on previously learned tasks. However, this did not affect the learning of new and advanced tasks. Overall, Leung et al. (2010) report that a multi-layered interface benefits older participants more than younger participants in terms of learning and time to complete the task.

6.3   SUMMARY OF RESEARCH OUTCOMES

Earlier studies by the authors have addressed some of these issues through two experiments. Experiment 1 systematically investigated redundancy (the use of both text and icons) in interface design as one of the strategies that could bridge the variability in older people's TF by accessing their knowledge of words or labels for functions, as well as icons that may relate to those functions, and helping them to use complex technological devices intuitively. Experiment 2 investigated the relationships between TF, the complexity of the interface (nested vs. flat), and intuitive use in older people. Overall, 137 participants between ages 18 and 83 years participated in these experiments. Experiments were designed based on those that had previously been used by the authors to investigate intuitive interaction (Blackler, Popovic, et al., 2010). TF, a variable that had been used previously by us and by others (e.g., O'Brien, 2010) was also measured in a similar way, and audiovisual data were coded using very similar heuristics to determine intuitive interaction.

In brief, in Experiment 1 (Reddy, Blackler, Popovic, & Mahar, 2010, 2011), contrary to what was hypothesized, older participants (65+) were significantly faster on the words-only interface when compared with the redundant and symbols-only interfaces. All age groups found the words-only interface more intuitive to use. Similarly, participants who had scored low on TF also took less time to complete the task on the words-only interface than on the redundant interface, which is not surprising, as TF was highly correlated with age. On the other hand, all TF groups and age groups took the most time to complete the tasks using the symbols-only interface. The most probable reason is that symbols are inherently ambiguous to interpret compared with words or text. Unlike verbal language, symbols do not have set syntactic and semantic rules that help to interpret them with certainty (Yvonne, 1989). Older people also made significantly fewer errors on the words-only interface compared with the redundant interface. Interestingly, there were no significant age differences between all three age groups in terms of errors made on the words-only interface. This is consistent with an earlier study that suggests that words-based interfaces produce fewer errors compared with symbols-only interfaces (Camacho, Steiner, & Berson, 1990).

Cognitive measures data from Experiment 1 also showed that visuospatial sketchpad capacity is affected most by aging and is also one of the most significant factors in the time taken to complete the task (Reddy et al., 2010). Both the redundant and symbols-based product interfaces used in trials were very visual in nature. This provides an explanation for older people taking more time on redundant and symbols-only interfaces. On a redundant interface, the amount of information to process is twice that of a words-only interface; this results in more stress on already limited visuospatial processing resources. On the other hand, age-related cognitive decline does not impact verbal processing and the age differences are minimal (Bäckman et al., 2001). This may be a reason why the performance of older people on the words-only interface was on a par with younger age groups in terms of errors. This is supported by an investigation by Jenkins, Myerson, Joerding, and Hale (2000), which suggested that aging adversely affects visuospatial information processing and memory for spatial information more than verbal information processing. O'Brien (2018, Chapter 5) also found evidence for older people using a more reflective, less intuitive approach that was slower but more accurate, suggesting a deliberate compromise in speed to achieve accuracy.

Overall, the findings of Experiment 1 support the literature that suggests that graphics-intensive interfaces can increase extraneous cognitive load and can hamper the learning of their functionality (Feinberg & Murphy, 2000; Sweller, 1994). Findings suggested that a simple text-based interface was much more helpful for older people than a redundant interface. Not only did older people use the text-based interface more intuitively, they also made fewer errors compared with their use of the redundant interface. This implies that they were able to learn the unfamiliar functions of the text-based interface during the task.

Experiment 2 investigated the relationship between age, TF, complexity in interface structure (nested vs. flat), and intuitive use (Reddy, Blackler, Popovic, & Mahar, 2013). Consistent with Experiment 1, most of the people who scored low on TF in Experiment 2 were from the older age group. TF scores of the older people were also more diverse compared with younger participants. As hypothesized, all age groups took significantly more time to complete the tasks on the nested interface when compared with the flat interface, probably because the nested interface needed more actions to complete the task (i.e., accessing the functions within the menus took more taps on the touchscreen). All age groups used the flat interface more intuitively compared with the nested interface, and age had a significant effect on time to complete the tasks on both types of interface. However, the differences between nested and flat interface use increased with age. This finding supports existing data that suggest older people find nested interfaces more difficult to use (Detweiler, Hess, & Ellis, 1996; Docampo Rama, 2001). Contrary to what was hypothesized, although older age groups made more errors, there were no significant differences in error rates between the use of the nested and flat interface types. The findings from this experiment also support processing speed theory (Salthouse, 2010), which suggests that older people tend to trade speed for accuracy, just as O'Brien (2018) also showed in observational studies.

Overall, older people scored lower on TF, took more time to complete the tasks, and used the interfaces less intuitively. However, the number of errors is one of the most crucial indicators for the successful use of a product interface. These data suggest that when the interface is designed with consideration of the cognitive limitations of older people (as both the nested and flat interfaces were), the differences in its use among age groups can be minimal. Apart from the oldest age group (73+), the differences in terms of intuitive use of the interface were also not significant. This supports research that suggests that working memory deficiencies in aging are mediated by coping mechanisms adopted by older individuals, especially when the task is simple (Brébion, Smith, & Ehrlich, 1997).

The outcomes of these two experiments have highlighted that older age groups, when compared with younger age groups, are very diverse in their capabilities in terms of TF and cognitive functioning, and that when the tasks are designed with consideration of the limitations of older people, the age differences are minimal for most age groups.

6.4 APPROACHES TO INTUITIVE INTERFACE DESIGN FOR OLDER PEOPLE

The findings from these two experiments suggest that, for a first-time encounter with a product interface, cognitive ability and prior experience are important factors that will decide the effectiveness of its intuitive use. This research also showed that, for novice users and users with low TF and age-related cognitive decline, the use of a simple text-based interface with flat interface structures would be beneficial. Text-based interfaces are easier to learn without external help for older people and for people with low TF and cognitive abilities. This finding is also supported by other related research that shows that a text-based interface aids the learning of an interface without external help (Camacho et al., 1990; Wiedenbeck, 1999).

However, text-based interfaces offer minimal spatial cues for visual searching (Cooper, Reimann, & Cronin, 2007). This could be a problem for older people, as age-related decline in visual information processing reduces their ability to search in a cluttered visual environment (Fozard & Gordon-Salant, 2001; Hawthorn, 2000, 2006). In contrast, symbols offer strong visual and spatial cues, and it is much easier to learn them and remember their location in the interface structure (Mertens, Koch-Körfges, & Schlick, 2011; Schröder & Ziefle, 2008b). Most of the research supports the view that a symbols-only interface is more efficient once users have learned the system (Camacho et al., 1990; Cooper et al., 2007; Mertens et al., 2011; Schröder & Ziefle, 2008a, 2008b; Yvonne, 1989). On the other hand, Experiment 1 showed that older people found the symbols-based interface the most difficult to use.

Widenbeck (1999) noticed that, on a redundant interface, people learned the meaning of symbols through their textual labels. Redundancy in the context of information communications is also often suggested for minimizing the ambiguity of symbols and for aiding their comprehension and learning (Wickens & Hollands, 2000; Wickens, Lee, Liu, & Becker, 2004). Keeping this in mind, it might be better that once the functions of the interface are learned on the text-based interface, it could progressively be switched to a redundant interface to help with learning the meaning of the associated symbols; finally, the interface could move to symbols only for the most effective use of the interface. Similarly, although a flat interface structure is good for making information more accessible, too many options tend to clutter the screen, resulting in more strain on the limited spatial ability of older people. Here again, it might be better to structure the options into a nested interface once the user learns all the functions of the system, thereby reducing visual clutter (Miller, 1981).

Kaufman and Weed (1998) noted that experience level may influence the effect of excess features on participants. Novices can be overwhelmed by lots of features within an interface, whereas experts will recognize interface elements more readily. McGrenere, Baecker, and Booth (2002) also agreed that novice users are able to accomplish tasks more accurately and rapidly with a simpler interface than the full version. McGrenere and Moore (2000) cited the training wheels interface—a real but simpler system for users to learn on. This was developed by Carroll (1990) and has been applied to various interfaces since. Novice users are able to complete tasks faster and with fewer errors using this type of reduced interface. This kind of approach was also recommended by Varland and Svensson (2006).

In practice, these findings strongly support the adaptable interface paradigm. The basic idea behind an adaptable interface is that a person can select the interface level based on their current ability and move up the levels as they acquire more experience with the system. One of the advantages of this approach is its flexibility to adapt to multiple user categories with varied prior experience and cognitive abilities. Most importantly, this flexibility allows users to develop their own learning curve based on their current abilities (Shneiderman, 2003). This process also supports Newell's design paradigm, which suggests that products be designed for older people first and then extended to use for other age groups. The basic idea is that, if a product is designed to accommodate, for example, people with low vision, it not only helps people with low vision but also people with normal vision in low-visibility conditions.

Until recently, implementation of adaptable interfaces was difficult as most consumer products had physical controls, and these are very difficult to customize to the abilities of the user. However, recent trends show that more and more products are using touchscreen interfaces with minimal or no physical controls. Therefore, it is now an appropriate time to embrace adaptable interface design. Moreover, recent research also suggests that, regardless of their cognitive or physical deficiencies, older people find touch-based products easier to learn and use (Häikiö et al., 2007; Isomursu, Häikiö, Wallin, & Ailisto, 2008; Taveira & Choi, 2009).

6.4.1 Intuitively Learnable Adaptable Interface System

Based on the literature and the outcomes of the two experiments described here, developing interfaces that are intuitive to learn rather than intuitive to use would be most practical. An intuitive-to-learn interface has the potential to counter the variability in TF and cognitive abilities over time and, eventually, to make the interface intuitive to use. An intuitively learnable interface in this context can be defined as an interface that allows a person to intuitively apply various strategies to learn and to successfully use a unique interface during first and early encounters.

Blackler, Desai, McEwan, Popovic, and Diefenbach (2018) have proposed an enhanced framework of intuitive interaction where all current ideas and approaches to intuitive interaction are mapped. The use of strategies such as image schemas, physical affordances, metaphors, and so on exploits basic and widespread knowledge of target users to address the variability in and even lack of specific familiarity with a domain/technology. These strategies are indeed very effective in designing intuitive interfaces by exploiting the prior experience users have of other domains. For example, the desktop metaphor has been used very effectively to design the file system of computer operating systems. However, most of these strategies are not as effective when it comes to implementing them for functions that do not have real-life equivalents, such as a Bluetooth option on mobile devices.

In general, for intuitive use, a user should be familiar with both the functions of an interface and how they are represented (Blackler, Mahar, & Popovic, 2010). The ILAIS framework addresses a lack of familiarity with (1) representations, by using language (instead of symbols/graphics), and (2) functions, by reducing complexity. In the context of the IUUI continuum of knowledge sources (Blackler & Hurtienne, 2007), the ILAIS framework exploits language as cultural knowledge for intuitive use during early encounters.

Figure 6.1 shows the theoretical structure of the multi-layered approach developed, based on our research, that facilitates people with diverse capabilities to intuitively learn an interface. Figure 6.1 also illustrates how levels of domain-specific prior experience are related to the levels of complexity of the interface structure, its functions, and their representations. A person with low prior experience and competence will find a flat interface structure with essential functionality and text-based

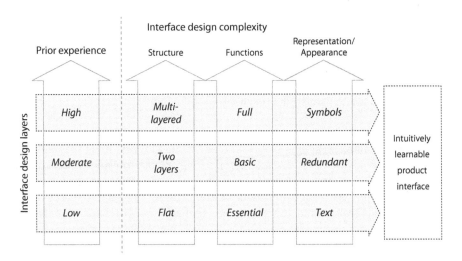

FIGURE 6.1 Multi-layered interface structure for ILAIS.

representation most beneficial for intuitive learning. On the other hand, a person with high prior experience and competence may find a multi-layered structure with full functionality and symbols-only representation more efficient to intuitively learn and use.

ILAIS generates interfaces based on the following two aspects:

1. *User ability*: This could be initiated by a user based on their own assessment or can be recommended by an intermediary, such as an evaluation application that can assess a user's TF with related technologies and their physical and cognitive capacities. This application could be built into the device as part of initial setup of a user profile.
2. *Customization rules*: These are developed by the designer for a particular device or application. Based on *customization rules*, ILAIS presents a user interface to match the current level of *user ability*. Customization provides rules that guide progressively increasing complexity of the interface based on the level of user ability. The directive that dictates the customization rules is to minimize the relearning of the interface as it progresses through different levels of complexity—for example, consistency in screen layout to minimize relearning when interface complexity is changed. Consistency is of course linked with prior experience and has been previously identified as essential for intuitive interaction (Blackler, Popovic, and Mahar, 2010). Elements are always added to the layout and never change their function. However, they can change from text to symbol representation, and they may move from a flat position to inside a nested menu.

6.4.2 IMPLEMENTATION

Figure 6.2 shows a simple example of how this framework can be implemented on a mobile device. In the example, there are three profiles:

1. *Low experience*: A single-layer, text-based minimal interface
2. *Moderate experience*: A single-layer, redundant interface with a conventional layout
3. *Experienced*: A symbol-dominant multi-layered interface

The next step would be to add multiple pages or nesting.

In terms of implementation platform, a device that allows direct manipulation or touch-based interaction is essential for allowing smooth transitions between the interface levels. The interface should be designed with flexibility to vary the complexity of structure from flat to nested, of functions from essential to full access, and of representation from textual to symbols based. Finally, issues related to implementation need further research to investigate different methods for changing interface levels in small increments. Whether the actual multi-layered interface implementation is adaptive, adaptable, or mixed initiative also needs further investigation to see how older people will cope with each.

FIGURE 6.2 Example of ILAIS implementation from text only to redundant.

6.5 CONCLUSION

In general, the findings from our two experiments and supporting literature show that older age groups are more heterogeneous in their TF and capabilities than younger people and that their use of contemporary technological devices is mediated by TF and cognitive abilities alongside chronological age. This highlights the fact that the development of entirely intuitive-to-use product interfaces from the word "go" is extremely difficult, mostly due to the diversity of older users' prior experience and cognitive abilities.

However, when interfaces are developed with consideration for the cognitive limitations and TF of older people, using simple text-based interfaces helps older people to learn the basic functions during early encounters with minimal errors. An adaptable interface allows older users and users with lower TF to use a simple text-based, flat-structured interface to help them learn and successfully use the new product interface during early encounters. Over time, and if required, the interface can progressively be changed to a symbols-based, nested interface for more efficient use. In other words, this framework will help to design products that are easy to learn and which, over time, are easy to use for older people with varied abilities. As we move toward software-driven touch-based interfaces, this adaptable interface framework removes the necessity for developing products for wider age and ability ranges.

ACKNOWLEDGMENTS

This research was supported by the Australian Research Council through Discovery Grant DP0877964.

REFERENCES

Akiki, P. A., Bandara, A. K., & Yu, Y. (2013). Preserving designer input on concrete user interfaces using constraints while maintaining adaptive behavior. Paper presented at the second workshop on context-aware adaption of service frontends (CASFE), CEUR (2013), London, UK.

Akiki, P. A., Bandara, A. K., & Yu, Y. (2014). Adaptive model-driven user interface development systems. *ACM Computing Surveys, 47*(1), 1–33.

Bäckman, L., Small, B., & Wahlin, A. (2001). Aging and memory: Cognitive and biological perspectives. In K.W. Schaie and S.L. Willis, S.L. *Handbook of the Psychology of Aging* (Vol. 5, pp. 349–377). New York, NY: Academic Press.

Bandura, A., Freeman, W., & Lightsey, R. (1999). Self-efficacy: The exercise of control. *Journal of Cognitive Psychotherapy, 13*(2), 158–166.

Blackler, A. (2008). *Intuitive Interaction with Complex Artefacts: Empirically-Based Research*. Saarbrücken, Germany: VDM.

Blackler, A., Desai, S., McEwan, M., Popovic, V., & Diefenbach, S. (2018). Perspectives on the nature of intuitive interaction. In A. Blackler (Ed.), *Intuitive Interaction: Research and Application* (pp. 19–40). Boca Raton, FL: CRC Press.

Blackler, A., & Hurtienne, J. (2007). Towards a unified view of intuitive interaction: Definitions, models and tools across the world. *MMI-Interaktiv, 13*, 37–55.

Blackler, A., Mahar, D., & Popovic, V. (2010). Older adults, interface experience and cognitive decline. Paper presented at OZCHI 2010, "Design—Interaction—Participation," 22nd Annual Conference on the Australian Computer–Human Interaction Special Interest Group, Brisbane, Australia.

Blackler, A., Popovic, V., & Mahar, D. (2010). Investigating users' intuitive interaction with complex artefacts. *Applied Ergonomics, 41*(1), 72–92.

Brébion, G., Smith, M., & Ehrlich, M. (1997). Working memory and aging: Deficit or strategy differences? *Aging, Neuropsychology, and Cognition, 4*(1), 58–73.

Camacho, M. J., Steiner, B. A., & Berson, B. L. (1990). Icons vs. alphanumerics in pilot-vehicle interfaces. *Proceedings of the Human Factors and Ergonomics Society Annual Meeting, 34*(1), 11–15.

Carroll, J. M. (1990). *The Nurnberg Funnel: Designing Minimalist Instruction for Practical Computer Skill*. Cambridge, MA: MIT Press.

Cooper, A., Reimann, R., & Cronin, D. (2007). *About Face 3: The Essential of Interface Design*. New York: John Wiley.

Czaja, S. J., & Lee, C. (2007). The impact of aging on access to technology. *Universal Access in the Information Society, 5*(4), 341–349.

Detweiler, M. G., Hess, S. M., & Ellis, R. D. (1996). The effects of display layout on keeping track of visuo-spatial information. In W. A. Rogers, A. D. Fisk, & N. Walker (Eds.), *Aging and Skilled Performance: Advances in Theory and Applications* (pp. 157–184). Mahwah, NJ: Lawrence Erlbaum.

Docampo Rama, M. (2001). *Technology generations handling complex user interfaces*. PhD thesis, Technische Universiteit Eindhoven, the Netherlands.

Docampo Rama, M., Ridder, H. D., & Bouma, H. (2001). Technology generation and age in using layered user interfaces. *Gerontechnology, 1*(1), 25–40.

Downton, A. C. (1993). *Engineering the Human–Computer Interface*. New York: McGraw-Hill.

Edmonds, E. (1981). Adaptive man–computer interfaces. *Computing Skills and the User Interface, 122*, 389–426.

Feinberg, S., & Murphy, M. (2000). Applying cognitive load theory to the design of web-based instruction. Paper presented at the 2000 IEEE Professional Communication Society International Professional Communication Conference and ACM Special Interest Group on Documentation Conference (IPCC/SIGDOC 2000), Cambridge, MA.

Findlater, L., & McGrenere, J. (2004). A comparison of static, adaptive, and adaptable menus. Paper presented at the SIGCHI conference on Human Factors in Computing Systems, Vienna, Austria.

Fozard, J. L., & Gordon-Salant, S. (2001). Changes in vision and hearing with aging. In *Handbook of the Psychology of Aging* (pp. 241–266).Elsevier; New York, NY.

Gajos, K. Z., Czerwinski, M., Tan, D. S., & Weld, D. S. (2006). Exploring the design space for adaptive graphical user interfaces. Paper presented at the Working Conference on Advanced Visual Interfaces, Venezia, Italy.

Gullà, F., Ceccacci, S., Germani, M., & Cavalieri, L. (2015). Design adaptable and adaptive user interfaces: A method to manage the information. In B. Andò, P. Siciliano, V. Marletta, & A. Monteriù (Eds.), *Ambient Assisted Living: Italian Forum 2014* (pp. 47–58). Cham, Switzerland: Springer.

Häikiö, J., Wallin, A., Isomursu, M., Ailisto, H., Matinmikko, T., & Huomo, T. (2007). Touch-based user interface for elderly users. Paper presented at the Ninth International Conference on Human–Computer Interaction with Mobile Devices and Services, Singapore.

Hawthorn, D. (2000). Possible implications of aging for interface designers. *Interacting with Computers, 12*, 507–528.

Hawthorn, D. (2006). *Designing Effective Interfaces for Older Users.* The University of Waikato, Waikato, New Zealand.

Hurtienne, J., Horn, A. M., & Langdon, P. M. (2010). Facets of prior experience and their impact on product usability for older users. In P. M. Langdon, P. J. Clarkson, & P. Robinson (Eds.), *Designing inclusive interactions* (pp. 123–132). London: Springer.

Hurtienne, J., Weber, K., & Blessing, L. (2008). Prior experience and intuitive use: Image schemas in user centred design. In P. Langdon, P. J. Clarkson, & P. Robinson (Eds.), *Designing Inclusive Futures* (pp. 107–116). London: Springer.

Isomursu, M., Häikiö, J., Wallin, A., & Ailisto, H. (2008). Experiences from a touch-based interaction and digitally enhanced meal-delivery service for the elderly. *Advances in Human–Computer Interaction, 2008*, 1–15 pp. doi:10.1155/2008/931701.

Jenkins, L., Myerson, J., Joerding, J. A., & Hale, S. (2000). Converging evidence that visuo-spatial cognition is more age-sensitive than verbal cognition. *Psychology and Aging, 15*(1), 157–175.

Kaufman, L. S., & Weed, B. (1998). Too much of a good thing? Identifying and resolving bloat in the user interface. *SIGCHI Bulletin, 30*(4), 46–47.

Langdon, P., Lewis, T., & Clarkson, J. (2007). The effects of prior experience on the use of consumer products. *Universal Access in the Information Society, 6*(2), 179–191.

Lavie, T., & Meyer, J. (2010). Benefits and costs of adaptive user interfaces. *International Journal of Human–Computer Studies, 68*(8), 508–524.

Lawry, S., Popovic, V., & Blackler, A. (2009). Investigating familiarity in older adults to facilitate intuitive interaction. Paper presented at IASDR 2009, Seoul, Korea.

Lawry, S., Popovic, V., & Blackler, A. (2010). Identifying familiarity in older and younger adults. Paper presented at DRS 2010, "Design and Complexity," Montreal, Canada.

Lawry, S., Popovic, V., & Blackler, A. (2011). Diversity in product familiarity across younger and older adults. Paper presented at IASDR 2011, Delft, the Netherlands.

Leonidis, A., Antona, M., & Stephanidis, C. (2012). Rapid prototyping of adaptable user interfaces. *International Journal of Human–Computer Interaction, 28*(4), 213–235.

Leung, R., Findlater, L., McGrenere, J., Graf, P., & Yang, J. (2010). Multi-layered interfaces to improve older adults' initial learnability of mobile applications. *ACM Transactions on Accessible Computing, 3*(1), 1–30.

Lewis, T., Langdon, P., & Clarkson, P. (2008). Prior experience of domestic microwave cooker interfaces: A user study. In P. Langdon and J. Clarkson (Ed.), *Designing Inclusive Futures.* London: Springer.

Lim, C. S. C. (2009). Designing inclusive ICT products for older users: Taking into account the technology generation effect. *Journal of Engineering Design, 21*(2–3), 189–206.

McGrenere, J., Baecker, R. M., & Booth, K. S. (2002). An evaluation of a multiple interface design solution for bloated software. *CHI Letters, 4*(1), 163–170.

McGrenere, J., & Moore, G. (2000). Are we all in the same bloat? Paper presented at the 2000 Graphics Interface Conference, Montreal, Canada. Online: http://www.dgp.toronto. edu/~joanna/papers/gi_2000_bloat.pdf.

Mertens, A., Koch-Körfges, D., & Schlick, C. M. (2011). Designing a user study to evaluate the feasibility of icons for the elderly. *Mensch und Computer 2011* (pp. 79–90). München: Oldenbourg.

Miller, D. P. (1981). Depth/breadth tradeoff in hierarchical computer menus. Paper presented at the 25th Annual Meeting of the Human Factors Society, Rochester, NY.

Naumann, A., Pohlmeyer, A., Husslein, S., Kindsmüller, M. C., Mohs, C., & Israel, J. H. (2008). Design for intuitive use: Beyond usability. Paper presented at CHI 2008, Florence, Italy.

O'Brien, M. (2018). Lessons on intuitive usage from everyday technology interactions among younger and older people In A. Blackler (Ed.), *Intuitive Interaction: Research and Application* (pp. 89–112). Boca Raton, FL: CRC Press.

O'Brien, M. A. (2010). *Understanding Human–Technology Interactions: The Role of Prior Experience and Age*. PhD thesis, Georgia Institute of Technology, Atlanta.

Peissner, M., Edlin-White, R. (2013). User Control in Adaptive User Interfaces for Accessibility. In: Kotzé P., Marsden G., Lindgaard G., Wesson J., Winckler M. (eds) Human–Computer Interaction: INTERACT 2013 Lecture Notes in Computer Science, vol 8117. Springer, Berlin, Heidelberg. doi: https://doi.org/10.1007/978-3-642-40483-2_44

Peissner, M., Häbe, D., Janssen, D., & Sellner, T. (2012). MyUI: Generating accessible user interfaces from multimodal design patterns. Paper presented at the Fourth ACM SIGCHI Symposium on Engineering Interactive Computing Systems, Copenhagen, Denmark.

Reddy, G. R., Blackler, A., Popovic, V., & Mahar, D. (2010). The effects of cognitive ageing on use of complex interfaces. Paper presented at OZCHI 2010, "Design—Interaction—Participation," 22nd Annual Conference on the Australian Computer–Human Interaction Special Interest Group, Brisbane, Australia.

Reddy, G. R., Blackler, A., Popovic, V., & Mahar, D. (2011). Ageing and use of complex product interfaces. Paper presented at IASDR 2011, "Diversity and Unity," Delft, the Netherlands.

Reddy, G. R., Blackler, A., Popovic, V., & Mahar, D. (2013). Intuitive use of complex interface structure, anxiety and older users. In P. K. et al. (Ed.), *Lecture Notes in Computer Science (LNCS)* (pp. 564–581). IFIP and Springer, Cape Town, South Africa.

Reddy, G. R., Blackler, A., Popovic, V., & Mahar, D. (2016). Designing for older people: But who is an older person? Paper presented at the Design Research Society 50th Anniversary Conference, Brighton, UK.

Salthouse, T. A. (2010). *Major Issues in Cognitive Aging*. Oxford, UK: Oxford University Press.

Schröder, S., & Ziefle, M. (2008a). Effects of icon concreteness and complexity on semantic transparency: Younger vs. older users. Paper presented at the 11th International Conference on Computers Helping People with Special Needs, Linz, Austria.

Schröder, S., & Ziefle, M. (2008b). Making a completely icon-based menu in mobile devices to become true: A user-centered design approach for its development. Paper presented at the 10th International Conference on Human–Computer Interaction with Mobile Devices and Services, Amsterdam, the Netherlands.

Shneiderman, B. (2003). Promoting universal usability with multi-layer interface design. Paper presented at the 2003 Conference on Universal Usability, Vancouver, Canada.

Shneiderman, B., & Plaisant, C. (2010). *Designing the User Interface: Strategies for Effective Human–Computer Interaction* (5th ed.). Pearson Education, Upper Saddle River, NJ.

Sloan, D., Atkinson, M. T., Machin, C., & Li, Y. (2010). The potential of adaptive interfaces as an accessibility aid for older web users. Paper presented at the 2010 International Cross Disciplinary Conference on Web Accessibility (W4A), Raleigh, NC.

Smirek, L., Zimmermann, G., & Beigl, M. (2016). Adaptive user interfaces as an approach for an accessible web of things. Paper presented at the Seventh International Workshop on the Web of Things, Stuttgart, Germany.

Stephanidis, C. (2001). Adaptive techniques for universal access. *User Modeling and User-Adapted Interaction, 11*(1), 159–179.

Stephanidis, C., Paramythis, A., Karagiannidis, C., & Savidis, A. (1997). Supporting interface adaptation: The AVANTI web-browser. Paper presented at the Third ERCIM Workshop on User Interfaces for All, Obernai, France.

Stuerzlinger, W., Chapuis, O., Phillips, D., & Roussel, N. (2006). User interface façades: Towards fully adaptable user interfaces. Paper presented at the 19th Annual ACM Symposium on User Interface Software and Technology, Montreux, Switzerland.

Sweller, J. (1994). Cognitive load theory, learning difficulty, and instructional design. *Learning and Instruction, 4*(295–312).

Taveira, A. D., & Choi, S. D. (2009). Review study of computer input devices and older users. *International Journal of Human–Computer Interaction, 25*(5), 455–474.

Varland, V., & Svensson, O. (2006). A minimalistic approach to designing graphical user interfaces. Unpublished manuscript, Game Systems and Interaction Research Laboratory (GSIL), School of Computing, Blekinge Institute of Technology, Blekinge, Sweden.

Wickens, C. D., & Hollands, J. G. (2000). *Engineering Psychology and Human Performance.* Prentice Hall, Upper Saddle River, NJ.

Wickens, C. D., Lee, J., Liu, Y., & Becker, S. G. (2004). *An Introduction to Human Factors Engineering* (2nd ed.). Prentice Hall, Upper Saddle River, NJ.

Wiedenbeck, S. (1999). The use of icons and labels in an end user application program: An empirical study of learning and retention. *Behaviour and Information Technology, 18*(2), 68–82.

Yvonne, R. (1989). Icons at the interface: Their usefulness. *Interacting with Computers, 1*(1), 105–117.

7 Development of the Technology Acceptance Intuitive Interaction Model

Dan McAran

CONTENTS

7.1 INTRODUCTION

There are two orientations that can be identified in intuitive interaction research. The first is research into the fundamental mechanisms of intuitive interaction with a technology and the consequent design tools and guidelines that can be created based on the results of this research (Blackler, 2006; Fischer, Itoh, & Inagaki, 2015; Hurtienne, 2009). The second orientation is focused on the perceptions of users in relation to technology and seeks to create instruments to measure the degree to which technology appears to the user as *intuitive* (Hurtienne & Blessing, 2007; Naumann & Hurtienne, 2010; Ullrich & Diefenbach, 2010). The research reported on in this chapter was an extension of the research focus on perceptions of a technology product as intuitive with the consequent creation of an *perceived intuitiveness* construct and measurement items.

In the last 25 years, user acceptance of technology theory based on the work of Davis (1986, 1989) has dramatically increased in explanatory power, yet there is concern that the theoretical developments that have occurred have not provided substantial interventions to promote user acceptance of technology in the workplace (Venkatesh, Davis, & Morris, 2007). The hypothesis for this research was that the creation of a *perceived intuitiveness* (PI) construct and the integration of the PI construct into the Technology Acceptance Model (TAM) would contribute to the resolution of this issue, which has been characterized as the "TAM Logjam" (Straub & Burton-Jones, 2007, p. 223).

The integration of the intuitive into technology acceptance has resulted in the creation of a new model, originally identified as the Technology Acceptance User Experience (TAUE) model (McAran, 2017), which included a PI construct but excluded the *perceived ease of use* (PEOU) construct. In this chapter, the model is re-identified as the Technology Acceptance Intuitive Interaction (TAII) model for reasons that will be detailed. The convergence of the concepts of *intuitiveness* and *ease of use* emerged as integral to this research, and additional evidence and literature are presented concerning the convergence of the concepts of intuitiveness and ease of use.

7.2 EXTANT THEORY

7.2.1 TECHNOLOGY ACCEPTANCE MODEL

TAM is an adaptation of the Theory of Reasoned Action (TRA) to user acceptance (Davis, 1986) and has been applied and replicated across a number of technologies (Venkatesh et al., 2007). TRA was created by Fishbein & Ajzen (1975) in order to formalize the meaning of *attitude* in relation to other constructs in a theoretical network: the concept of attitude was prevalent in social psychology but was not theoretically well developed (Sheppard, Hartwick, & Warshaw, 1988). Davis (1989) first presented the TAM model using three factors: perceived usefulness (PU), PEOU, and attitude; *self-reported use* and *behavioral expectation* were included as the dependent variables. Attitude was subsequently dropped from the TAM model (Taylor & Todd, 1995).

Benbasat & Barki (2007) argued that the psychological, social, and technological factors central to technology acceptance have evolved, such that PEOU and PU are no longer the sole important factors influencing technology acceptance. A person will normally hold only a small number of salient beliefs (Ajzen, 2012; Fishbein & Ajzen, 1975). In the present research, it was postulated that the PI of the technology product was a salient belief affecting the acceptance and use of a computer technology product.

There is concern (most profoundly expressed by several authors in the 2007 special issue of the *Journal of the Association for Information Systems* entitled "Quo Vadis TAM: Issues and Reflections on Technology Acceptance Research" [Hirschheim, 2007]) that, despite the extent of the research performed on user acceptance of technology, few design specifications or interventions have emerged to enhance or promote user acceptance of technology. Hevnerm, March, & Park (2004) see IS as partially a "design science" (p. 76), and there is an increasing focus on design in *information science* (IS) research (Gregor, 2006). The design of intuitive technology is an emergent area of research, and integrating the PI construct into the TAM model addressed the call to link technology acceptance to technology design (Venkatesh & Bala, 2008).

7.2.2 TAM Constructs

As outlined by Chuttur (2009), the original TAM model of Davis (1986, 1989) has been extended, with additional constructs that mainly function as antecedents of PEOU and PU. Examples of such antecedent constructs are *compatibility* (COM), *computer self-efficacy, computer anxiety,* and *perceived enjoyment* (Chau & Hu, 2002; Venkatesh, Morris, Davis, & Davis, 2003). The present research was consistent with the method used by Chau & Hu (2002) and Venkatesh et al. (2003) in that it integrated the novel PI construct into the TAM model. This section will briefly describe the TAM constructs relevant to this chapter, starting with a definition of each.

- COM is defined as the degree to which the use of technology is perceived by the user to be consistent with his or her practice style or preferences (adapted from Chau & Hu, 2002).
- PU is defined as the degree to which an individual believes that using a particular system will enhance his or her performance (Davis, 1986, p. 26).
- *Degree of voluntary use* (VOL) is defined as the degree to which the use of the technology is perceived as being voluntary or of free will (Moore & Benbasat, 1991, p. 195).
- PEOU is defined as the degree to which an individual believes that using a particular system will be free of physical and mental effort (Davis, 1986, p. 26).
- PI is defined as the degree to which the technology product is perceived by the user as capable of being used without conscious awareness of rational thinking (adapted from Shirley & Langan-Fox, 1996, p. 564).

7.2.2.1 Compatibility

COM was one the most significant factors in Diffusion of Innovation Theory (Rogers, 1983). Tornatzky & Klein (1982) found COM to be the most studied attribute ahead of *relative advantage* (adapted by Davis as PU) and *complexity* (adapted by Davis as PEOU). Moore & Benbasat (1991) and Agarwal & Prasad (1997) also highlight the importance of COM. Chau & Hu (2002) added COM to the Technology Acceptance by Individual Professionals (TAIP) model they developed to study the acceptance of telemedicine technology by Hong Kong physicians, as it was their view that physicians would find technology that fit with their practice style and preference desirable. Chau & Hu (2002) found the COM to PU path to be the most important in their model and emphasized the importance of COM in the acceptance of telemedicine technology by Hong Kong physicians.

Van Ittersum, Rogers, Capar, Caine, O'Brien, Parsons, & Fisk (2006) performed an extensive review of the technology acceptance literature; they found COM to be key to technology acceptance. Fischer, Oelkers, Fierro, Itoh, & White (2015) identify two components of knowledge related to user knowledge of computer interfaces: (1) compatible features with which the user has prior knowledge and (2) innovative features that require the creation of new knowledge. They deduce compatible features with which the user has prior knowledge as being intuitive. This finding relating COM to intuitiveness is also supported by the research of Blackler & Hurtienne (2007). Consequently, COM, which can be related directly to intuitive use, was included as one of the constructs in the research model designed to investigate the acceptance of technology used by legal professionals.

7.2.2.2   Perceived Usefulness

Davis (1986) created the PU construct as we know it in his doctoral dissertation. Using the literature of management information systems (MIS), lab experiments, field studies, and human factors literature, Davis integrated PU into the TAM model. PU has proved the most important of predictive variables in TAM; King & He (2006) characterize its effect as "profound" (p. 751). In the now prominent Unified Theory of Acceptance and Use of Technology (UTAUT) model, which is an extension of TAM, *performance expectancy* and *effort expectancy* correspond to PU and PEOU in the TAM model, respectively (Venkatesh et al., 2003). The concern has been expressed that the high level of research on the relationship of PU to technology acceptance has distracted researchers from the importance of design in relation to the information technology artifact (Benbasat & Barki, 2007).

7.2.2.3   Degree of Voluntary Use

Technology acceptance research using TAM is based on TRA, and a central postulate of TRA is that performance is voluntary (Sheppard et al., 1988). Kroenung & Eckhardt (2015) performed a meta-analysis of 119 articles from 14 top IS journals to identify significant factors that influence the attitude–behavior relationship, and one of the factors identified as having a significant effect was voluntariness. The UTAUT model (Venkatesh et al., 2003), which is also an adaptation of TAM, shows *voluntariness of use* acting through *subjective norm* (SN), which is defined as the perception by the user that people who are important to the user think the user should use the technology product. In this research, to evaluate the effect of VOL in the absence of SN, VOL was hypothesized to have a moderating effect on the paths from the exogenous variables to the endogenous variable: namely, the paths from PI, PEOU, PU, and COM to USE. In addition, consistent with the non-binary nature of voluntary use identified by Moore & Benbasat (1991), VOL was measured in this research using a slide bar with a 1–100 scale.

7.2.2.4 Perceived Ease of Use

In a pre-test of PEOU measurement items, Davis (1986, p. 86–87) clustered the responses received into three categories: (1) physical effort, (2) mental effort, and (3) easy to learn. Davis dropped three items related to error recovery, unexpected behavior, and error proneness that did not cluster into these three categories and received low-priority ratings by participants. The dropped items correspond to the elements *feedforward methods* and *lenient learning environments* that form part of the O'Brien, Roger, & Fisk (2010) definition of intuitive human–computer interaction. An additional item (*provides guidance*) was also dropped. The significant reduction of measurement items for the TAM PEOU construct may have prevented the full capture of the domain content of PEOU related to intuitive interaction. This may also be the partial cause for the relatively weak explanatory power of PEOU (Davis, 1986).

The limited domain content of PEOU is additionally evident in reviewing the measurement items of PEOU utilized in the development of Technology Acceptance Model 3 (TAM3), developed by Venkatesh & Bala (2008), in that the concepts *adapts, feedforward*, and *lenient learning environment* cannot be identified in the domain content of the PEOU items. The research of Chau & Hu (2002) on the use of telemedicine technology by physicians suggested that the total effect on PEOU on behavioral intention (BI) was small (p. 211). From this finding, it can be inferred that for technology used by professionals—in Chau & Hu's case, physicians in their professional practice—the PEOU of technology has little influence on whether the technology will be used.

7.2.3 INTUITIVE INTERACTION RESEARCH AND THE PI CONSTRUCT

It can be postulated that the user acceptance of technology is a design problem and, as such, this research looked to the design of technology perceived as intuitive as a potential solution to the issue of technology acceptance. Consequently, the work done on intuitive interaction design was of paramount importance in this research. Intuitive interaction with technology is a phenomenon that has been theorized and empirically validated by a number of researchers with varying research approaches (Blackler, 2006; Blackler & Hurtienne, 2007; Fischer et al., 2015; Hurtienne, 2009; Hurtienne & Blessing, 2007; Naumann & Hurtienne, 2010; O'Brien et al., 2010; Ullrich & Diefenbach, 2010). *Intuitive* in relation to technology means more than familiar or easy to use or understand (O'Brien et al., 2010, p. 7), and Blackler, Popovic, & Mahar (2010) see intuitive interaction as also being related to unconscious or implicit knowing (p. 13). PI was developed independently as a construct in this research, but there have been other references to PI in the literature.

Ullrich & Diefenbach (2010), when developing their INTUI questionnaire of intuitive interaction, added to their *effortless, gut feeling, verbalizability*, and *magical experience* components an additional measure to evaluate the overall PI of the interaction using a seven-point scale (p. 803). In a subsequent paper, Diefenbach & Ullrich (2015) identify the result of intuitive interaction as PI but do not identify the results of their research as creating a PI construct, supporting the gradual emergence of PI as a construct. Similarly, Islam (2014), in his doctoral dissertation researching intuitive interface signs (e.g., icons and navigational links) from a semiotic theory perspective, sought to measure the PI of interface signs using a PI score. McEwan, Blackler, Johnson, & Wyeth (2014), in their study of intuitive interaction with video games, used the three-item Intuitive Controls component of the Player Experience of Need Satisfaction (PENS) instrument. Macaranas, Antle, & Riecke (2015) used a seven-level score of PI in their evaluation of intuitive interaction with a whole-body system. Recently, Bakke (2015) related PI to *flow* and analyzed intuitiveness from an Activity Theory approach.

The following definition of intuitive human–computer interaction from O'Brien et al. (2010) was used as a working definition in this research: "interactions between humans and high technology in lenient learning environments that allow the human to use a combination of prior experience and feedforward methods to achieve their functional and abstract goals" (p. 107). Feedforward is "the modification or control of a process using its anticipated results or effects" (Oxford Online Dictionary, 2017). A practical example of feedforward is smartphone technology that uses feedforward methods as part of predictive text technology. The concept of feedforward has been identified in several fields (e.g., control engineering; Gorinevsky, 2003) and has been found to improve performance.

Blackler & Popovic (2015) identify the importance of intuitive interaction to practitioners, and the degree to which technology is intuitive is emerging as a design standard. This can be seen in an excerpt from Shaw (2011), where a physician comments on the use of electronic health records in practice:

> It hasn't changed the way I worked. ... It's very intuitive as far as the way the family physicians practice. Everything on the screen is—it just automatically fits in. The system works exactly the way I work. That's the beauty of that particular system.
>
> **(p. 159)**

The importance to practitioners of intuitive technology can also be seen in the Apple design guidelines (Apple, 2017a, 2017b). In the "Apple iOS Human Interface Guidelines," advice is provided to create interfaces that are intuitive (Apple, 2017a). In the "macOS Human Interface Guidelines," creating intuitive apps is also highlighted (Apple, 2017b). Because of the importance of intuitive interaction in technology design, the PI construct was hypothesized to have a positive effect on all constructs in the research model.

7.3   EVALUATION OF THE PI CONSTRUCT IN THE TAM MODEL

The objective was to integrate a novel PI construct into the TAM model. The target population was legal professionals (lawyers, paralegals, law clerks, legal assistants) who used social media, online groups, or services related to legal technology. The research was limited to legal technology. *Legal technology* is a summary term that is used to mean technology specifically designed for the legal profession and used by legal professionals to perform legal work.

To enable use of partial least squares structural equation modeling (PLS-SEM), a minimum number of 60 usable responses was required for the creation of a PLS-SEM model with adequate statistical power (Hair, Hult, Ringle, & Sarstedt, 2014, p. 21). The following hypotheses sought to evaluate the research model (Figure 7.1).

7.3.1 Hypotheses

The following hypotheses for this research were derived from the literature.

- H1: PI is positively related to PU.
- H2: PI is positively related to PEOU.
- H3: PI is positively related to COM.
- H4: PI is positively related to the combined measure of USE (consisting of *degree of use* and *degree of feature use*).
- H5: The degree to which the technology use is voluntary will have a significant moderating effect on the paths from PI, PU, PEOU, and COM to USE.

The unit of analysis was the individual legal technology user.

7.3.2 Research Design

The research TAM model is shown in Figure 7.1. This model was the TAM model of Davis (1986) with the addition of the COM construct from Chau & Hu (2002), which itself was adapted from Moore & Benbasat (1991), and the proposed PI construct. COM was included because of its prominence in technology acceptance research (Chau & Hu, 2002; Moore & Benbasat, 1991; Tornatzky & Klein, 1982). The proposed PI construct was hypothesized as having a positive effect on all the existing constructs in TAM and the COM construct; PI is shown as having a direct effect on paths from PU, PEOU, and COM to USE. The advantage of this model is that it is simple but still likely captures most of the significant relationships (McAran, 2017).

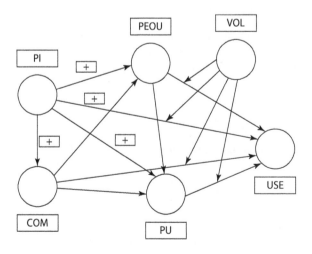

FIGURE 7.1 Original research model. COM – Compatibility; PEOU – Perceived Ease of Use; PI – Perceived Intuitiveness; PU – Perceived Usefulness; VOL – Degree of Voluntary Use.

Recruitment was conducted in the 8-month period from October 2014 to May 2015 using the following methods: (1) posts to LinkedIn (www.linkedin.com) social media legal groups; (2) the LinkedIn InMail service, with InMails sent to 958 LinkedIn members; and (3) emails to 180 legal professionals who had posted on Technolawyer (www.technolawyer.com). A link to the Qualtrics (2015) online questionnaire was provided in the posts and emails sent. Total responses received to the Qualtrics online survey were 218 with 154 usable responses; the response rate was 19.15%.

7.3.3   DEPENDENT VARIABLE

In this research, as is common in technology acceptance research, self-reported use was employed with a selection of six commonly used legal technology products as the basis of responses. Respondents were able to also "write in" a legal technology product.

Actual use is the preferred measure of use in technology acceptance research (Venkatesh et al., 2003; Straub, Limayem, & Karahanna-Evaristo, 1995; Wu & Du, 2012). The dependent variable USE was measured using two reflective indicators: degree of use and degree of feature use. These were both measured using a slide bar (sliding) scale allowing for responses from 1 to 100 as an indication of the degree of use (Hair, Celsi, Money, Samouel, & Page 2016). A slide bar scale was also used to measure VOL, as explained previously. USE within the single instrument was measured first, whereas the exogenous constructs PI, PEOU, PU, and COM were measured subsequently using seven-point Likert scales with ratings from "Strongly agree" to "Strongly disagree." Consequently, the risk of common method bias (CMB) may be reduced (Podsakoff, MacKenzie, Lee, & Podsakoff, 2003).

7.3.4   QUANTITATIVE PILOT STUDY

The quantitative pilot study used an online Qualtrics survey and consisted partly of paid Qualtrics data. In addition, respondents were solicited using LinkedIn social media legal profession–related groups, the TechnoLawyer (www.technolawyer.com) email news service, and personal contacts. A total of 131 responses were received by the web-based Qualtrics pilot study. The survey was conducted from April to August 2014. A large pilot study was undertaken to reduce the number of measurement items of PI, to identify the most common legal technology products used, and to determine the best method of soliciting respondents. After review, 74 (56.49%) responses were identified as usable.

There were 10 measurement items developed from the literature for the PI construct. The Questionnaire for the Subjective Consequences of Intuitive Use (QUESI) instrument (Naumann & Hurtienne, 2010) was also identified, and 13 of the QUESI measurement items were reformatted and included among the measurement items for PI in the quantitative pilot study, for a total of 23 measurement items. Based on the quantitative pilot study, the 23 original measurement items developed for the PI construct were reduced to 15 items. Further, based on the pilot study results, the decision was made to list six commonly used legal technology products as the basis of the responses.

7.4 RESULTS

It was decided to perform the primary analysis of the results on a subset of the data that consisted of responses concerning legal research technology (henceforth identified as the Westlaw data set). Legal research technology is used to research legislation and legal cases. Although these products have different designs, they all perform similar functions and provide the same types of results, consequently forming a homogeneous set of technologies. The remaining segment of the data was identified as the non-Westlaw data set.

PLS-SEM is increasingly used in social science research, particularly accounting, marketing, and IS research, and is particularly useful for exploratory research (Hair et al., 2014). In IS research, PLS-SEM has been frequently used to analyze the results of TAM and related models. The focus of PLS-SEM is to maximize the amount of variance explained. The results were analyzed using PLS-SEM in SmartPLS 3 (Ringle, Wende, & Becker, 2015). In research using PLS-SEM, reporting a standard set of analyses is recommended (Hair et al., 2014). These analyses are grouped under the headings *measurement model*, which reports on the relationships between the measurement items and the latent constructs, and *structural model*, which reports on the relationships between the latent constructs themselves. These detailed results can be found in McAran (2017). CMB is variance related to the measurement method used rather than measurement items representing the constructs (Podsakoff et al., 2003). Tests for CMB and non-response bias indicated they were not of concern.

The results showed colinearity between the PEOU, PI, and COM constructs. Based on these indications of colinearity, it was decided to reorganize the model using a second-order construct as recommended by Hair et al. (2014), identified as *intuitive interaction* (II). The second-order II construct was created from the existing first-order reflective constructs (PEOU, PI, and COM). The revised model (Figure 7.2) showed only positive paths and had acceptable R^2. In addition, bootstrap analysis showed all paths as significant, except for the path for the moderator variable VOL to USE and the interaction term for VOL on the path PU to USE.

The next step was the evaluation of the effect size (f^2) of this revised model. An attempt was also made to evaluate the f^2 effect of removing each of the component first-order constructs of II. This procedure gave unexpected results: removing PEOU resulted in an increase in R^2; the consequent f^2 was negative, $-.0109$. A similar analysis was performed in regards to the predictive relevance, Q^2 and q^2, and showed similar results. Removing PEOU from the model increased the explanatory ability of the model (increasing R^2) and made the model more parsimonious. The resulting model, now identified at the TAII model, is shown in Figure 7.3. All of the remaining PLS-SEM analysis is presented in regards to the TAII model (Figure 7.3) only.

The analysis performed for the TAII measurement model consisted of indicator reliabilities (*Cronbach's alpha* and *composite reliability*), *Fornell–Larcker criteria*, and *average value extracted*. Details of the measurement model analysis

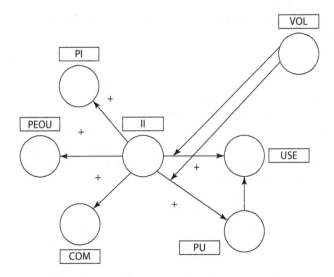

FIGURE 7.2 First II second-level construct model. COM – Compatibility; PEOU – Perceived Ease of Use; PI – Perceived Intuitiveness; PU – Perceived Usefulness; VOL – Degree of Voluntary Use; II – Intuitive Interaction.

can be found in McAran (2017). The analysis performed for the TAII structural model consisted of R^2 values, bootstrap results, effect size (f^2), and predictive relevance (Q^2 and q^2). Details of the structural model analysis can be found in McAran (2017). Figures 7.4 and 7.5 provide summary information on the structural model results.

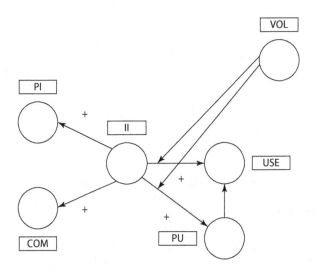

FIGURE 7.3 Final model: TAII model. COM – Compatibility; PEOU – Perceived Ease of Use; PI – Perceived Intuitiveness; PU – Perceived Usefulness; II – Intuitive Interaction; VOL – Degree of Voluntary Use.

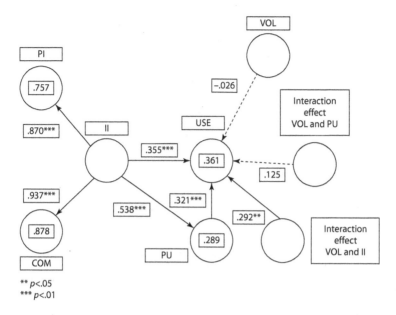

FIGURE 7.4   Structural model paths: TAII model. COM – Compatibility; PEOU – Perceived Ease of Use; PI – Perceived Intuitiveness; PU – Perceived Usefulness; VOL – Degree of Voluntary Use; II – Intuitive Interaction.

7.4.1   COMPARATIVE RESULTS OF TAII AND TAM

As mentioned, it was decided to perform the primary analysis of the results on a subset of the data that consisted of responses concerning legal research technology (the Westlaw data set): a homogeneous set. All the data consisted of the Westlaw data set ($n=94$) and the non-Westlaw data set ($n=60$). Table 7.1 compares the results obtained for the Westlaw data set and all data using the TAII model and the TAM model (Davis, 1986).

7.5   DISCUSSION

7.5.1   SUMMARY OF FINDINGS

Investigating the original research model showed high latent variable correlations for COM, PI, and PEOU. When PEOU was removed from the model, the new model, now identified as the TAII model, showed reasonable R^2, and the bootstrap showed all paths as significant, except for VOL to USE and one interaction term for VOL. The R^2 found for the TAII model using the Westlaw data set ($n=94$) was 0.3612, which is similar to the overall level of explanation found in IS research (Burton-Jones & Straub, 2006, p. 230).

The hypotheses for this research were addressed as follows:

- H1: PI will have a positive effect on PU. In the TAII (final model), the II construct, which consists of PI and COM as first-order reflective constructs, was found to have a positive effect on PU. This hypothesis was supported.

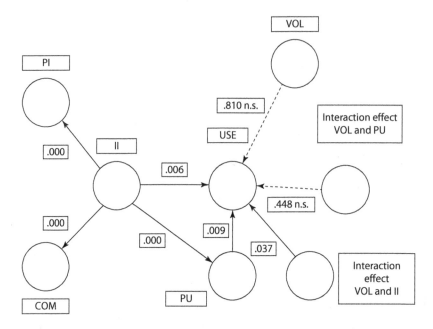

FIGURE 7.5   Structural model bootstrap: TAII model. Dashed lines indicate non-significant paths. COM – Compatibility; PI – Perceived Intuitiveness; PU – Perceived Usefulness; II – Intuitive Interaction; VOL – Degree of Voluntary Use.

- H2: PI will have a positive effect on PEOU. In the TAII (final model), PEOU was not present as a construct. This hypothesis was not supported.
- H3: PI will have a positive effect on COM. In the TAII (final model), COM was combined with PI as a second-order construct identified as II. This hypothesis was not supported.
- H4: PI will have a positive effect on the combined measure of USE. In the TAII (final model), the II construct, which consists of PI and COM as first-order reflective constructs, was found to have a positive effect on the combined measure of USE. This hypothesis was supported.
- H5: The degree to which the technology use is voluntary will have a moderating effect on the paths from PI, PU, PEOU, to the combined measures of USE. In the TAII (final model), VOL did not have a significant effect on the paths to the combined measures of USE. In addition, the interaction term for VOL on the path from PU to the combined measures of USE was not significant. The interaction term for VOL on the path from II to the combined measures of USE was significant. The hypothesis was partially supported.

7.5.2   Convergence of Ease of Use and Intuitiveness

In the TAII model, PI has replaced PEOU. *Convergent validity* measures whether a construct correlates with other constructs that also measure the same characteristic.

TABLE 7.1
Comparative Results: TAII and TAM Models

Data Set	TAII	TAM	Comments
Westlaw[a] data set ($n=94$)	R^2 .3612	R^2 .3063	The TAII model had
	Bootstrap paths	*Bootstrap paths*	better R^2 than the
	significant: ($p<.01$)	*significant:* ($p<.01$)	TAM model.
	II to USE	PEOU to PU	In the TAM model, the
	II to PU	PU to USE	path PEOU to USE
	PU to USE	*Bootstrap path not*	was not significant.
		significant:	
		PEOU to USE	
		$p=.1935$	
All data[a] ($n=154$)	R^2 .2731	R^2 .2145	The TAII model had
	Bootstrap paths	*Bootstrap paths*	better R^2 than the
	significant: ($p<.01$)	*significant:* ($p<.01$)	TAM model.
	II to USE	PU to USE	In the TAM model, the
	II to PU	PEOU to PU	path PEOU to USE
	Bootstrap paths	*Bootstrap paths not*	was not significant.
	significant: ($p<.05$)	*significant:* PEOU to	
	PU to USE	USE	
		$p=.1557$	

[a] In this comparison of the models, the same three measurement items for PI are used.

Discriminant validity measures a construct's correlation with constructs that do not measure the specified characteristic (Fink, 2008). The results support the emerging convergence of the concepts of intuitiveness and ease of use, and support is also found in the literature. A search of the recent Vermesan & Frie (2013) text for "ease of use" returned no results, indicating the declining importance of the concept. Kappelgaard & Bala (2011) investigated intuitiveness and ease of use at the same time for three growth hormone injection devices. They also relate intuitive interaction to minimal training. Questionnaires were used to measure intuitiveness and ease of use: participants rated one device as the most intuitive and easiest to use. Marchal, Moerman, Casiez, & Roussel (2013), in their paper on multi-touch three-dimensional navigation techniques, mention intuitiveness and ease of use as jointly desirable design objectives; notably, they did not delineate any differences between these two concepts. A similar joint evaluation of intuitiveness and ease of use can be found in Lohmann, Negru, & Bold (2014), who performed a comparative evaluation of the ProtégéVOWL plugin for ontology representations used in the Semantic Web with other available plugins for the Protégé ontology editor. Notably, Petter, DeLone, & McLean (2013) included ease of use and intuitiveness together as examples of the System Quality IS success factor.

Writing on the emergent Internet of Things (IoT), Fantana et al. (2013) comment, "Simple, intuitive use and (almost) self-explaining are important for the overall IoT application acceptance. The IoT application should ideally be context aware and adapt to the skills of the user and location or environment aspects" (p. 159). The references to technologies that are "self-explaining, context aware, and adapt to the skills of the user and location or environment aspects" are similar to the three measurement items identified for the PI first-level reflective construct associated with the second-level II construct.

- *I find this legal technology product can be used in my practice with minimal training.*

  This item is closely related to *self-explaining* (Fantana et. al., 2013) in that it can be expected that technology that is self-explaining will consequently require minimal training.

- *When I use this legal technology product in my practice, it is always clear to me what I have to do to use the product.*

  This item is closely related to *self-explaining* and *context aware* (Fantana et al., 2013), in that it can be expected that technology that is self-explaining and context aware will consequently result in a situation where the user is always clear what they have to do to use product.

- *When I use it in my practice, this legal technology product adapts to my specific goals as I enter responses.*

  This item is closely related to "adapt to the skills of the user and location or environment aspects" (Fantana et al., 2013), in that it can be expected that technology that adapts to goals of the user would also potentially adapt to other related factors such as the user's environment and location.

PEOU was dropped from the TAII model as it added no additional explanatory power. This can be related to the research of Chau & Hu (2002). In the TAIP model that they used to study the acceptance of telemedicine technology by Hong Kong physicians, they found that the total effect of PEOU on behavioral intention to use the telemedicine technology (the dependent variable in their research) was low, with a total path effect of PEOU to BI of .05.

It is notable when the terms *perceived ease of use, intuitive interaction, perceived intuitiveness*, and *intuitiveness* are first found, and the frequency with which they are found, in academic discourse. Google Scholar searches were performed for each of these terms for each year from 1980 to 2016. References to PEOU started to appear in significant numbers around the time that Davis (1986) published his thesis outlining TAM; prior to this, there were only a small number of references to PEOU to be found in academic literature. There were 34 references to PEOU appearing in Google Scholar in 1986, and 24 of these were references to TAM or Davis's work. In 1989 Davis published the first paper outlining TAM in *Management Information Systems Quarterly (MISQ)*. In the period 1989–2002, references to PEOU grew to 609. In 2003, UTAUT (Venkatesh et al., 2003) was published, which greatly increased interest in technology acceptance research, resulting in 820 references to PEOU. This number increased to 7300 per year in 2016, reflecting the degree of acceptance of TAM and UTAUT across diverse academic disciplines.

A search was also conducted using the term "intuitive interaction" using the Google Scholar search tool. In the period 1961–1979, there were 16 references to II; however, none of these referred to technology (and prior to 1961, there were no such references). In the period 1980–1987, there were a total of 12 references to II; again, none related to technology. In 1984, the Macintosh computer was introduced. In 1985, Windows 1.0 was released. In 1988, the first references to II appear related to technology. Sutcliffe (1988), in *Human–Computer Interface Design*, made reference to "the intuitive model of interaction" (p. 168). In her master's thesis, Lambert (1988) also referenced "intuitive interaction" (p. 17) in regards to computer-aided instruction. In 1989, 1990, and 1991, there were five references to II related to technology. In 1992 there were 4 such references, out of 13 in total, related to computer technology. Notably, Bodi and Kaulichb (1992) specifically discussed "intuitive user interfaces."

In 1993, 12 of 25 references to II related to computer technology. References to II increase after 1993. In 2002, the year of the first paper by Blackler, Popovic, & Mahar (2002) relating to intuitive use, there were 142 corresponding search results appearing in Google Scholar for II. In the period 2003–2016, there was a steady increase in these references; in 2016, there were 614 such results.

There are, at this point, only limited references to "perceived intuitiveness" appearing in academic literature. A Google Scholar search (1980–2016) returned 84 uses of this term. The first reference to PI in the period 1980–2016 was in 1990; this reference, however, was concerned with Husserl's philosophy. In 2001 and 2002, there were single references to a nurse's PI in regards to patients. In 2003, there were four references, one again related to nursing practice, but three are identifiable as being related to technology. References to PI increased alongside the launch of the Apple iPhone in 2007: there were 5 such references in 2007; 5 in 2009; 6 in 2010; 9 in 2011; 7 in 2012; 7 in 2013; 13 in 2014; 11 in 2015; and 9 in 2016. All but one of the 2016 references to PI related to technology—mainly computer-related technology.

References to "intuitiveness," as distinct from "perceived intuitiveness" or "intuitive interaction," show a different pattern. There was a steady number of references to intuitiveness in the period from 1980 to the early 1990s, averaging about 50 per year; however, few of these related to technology. Interestingly, in 1980, the term "intuitiveness" appeared in relation to an MIT master's thesis concerning input devices for computer technology (Schmandt 1980). There was a noticeable increase in references to intuitiveness after Raskin's (1994) article entitled "Intuitive Equals Familiar." In 1994, there were 195 such references, which increase to 1330 in 2007, the year of the introduction of the Apple iPhone, after which such references increased steadily until 2016, when they reached 2670.

A similar finding in regards to references related to *intuitive technology* (IT) was found by McAran & Manwani (2013, 2014) using articles in issues of *MISQ* from the commencement of publishing in 1977 to the end of 2012. A similar review was made for *Information Systems Research (ISR)*, starting with the commencement of publication to the end of 2009 (McAran & Manwani, 2013, 2014). The PDF versions of *MISQ* and *ISR* articles were searched three times for the text "intu" in order to identify the word "intuition" and related terms. Codes were assigned to text using a method consistent with Grounded Theory (Glaser & Straus, 2009) and were reviewed several times. The process of code generation resulted in 21 codes; details of this research can be found in McAran & Manwani (2013, 2014). Of the 986 codes assigned, 44 were assigned to the IT code, representing 4.5% of total codes assigned; these codes were analyzed over time and by publication, as shown in Table 7.2. As can be seen, the number of IT codes assigned has been fairly consistent in the range 0.4–0.8 codes per year/per publication in the period 1977–2009. However, for *MISQ* in the period 2010–2013, there was a significant increase to 4.3 codes per year; this time period can then be identified with the emergence of the concept of intuitive technology in the mainstream. Again, there would be general correspondence to the emergence of smartphone technology.

It is noteworthy that TAM was developed by Davis (1986) in his doctoral thesis, with the first article introducing TAM published in *MISQ* in 1989. As can be seen in Table 7.2, the main emergence of the concept of intuitive technology in IS research occurred after this time. This explains, in part, why the concept of the intuitive was not part of the original formation of TAM by Davis.

The use of these terms is important, because they represent categories that provide basic information about human thought processes (Lakoff, 1987). In this circumstance, intuitive technology and the intuitiveness of technology can be viewed as emergent human concepts represented by the use of these words. We are seeing an evolution of the language related to technology acceptance and, by inference, the mental models related to technology acceptance. As has been indicated, this can also be seen in academic articles using the terms *intuitiveness* and *ease of use* in a related manner. The near convergence of these categories can also be seen in the work of Kovanović et al. (2017), who evaluated a novel software tool designed to manage data for the Massive Open Online Courses (MOOC) online learning method. In their user evaluation of the MOOC software tool, they created an instrument based on TAM (Davis, 1986). Two components of the software tool were evaluated: (1) the Dashboard screen, and (2) the Educational Intervention screen. Two of the four items for the Dashboard measure of PEOU refer to "intuitive." For the Educational Intervention component of the software tool, two out of three items refer to "intuitive." This provides additional evidence of the increasing convergence of the categories of phenomena that both ease of use and intuitiveness represent.

TABLE 7.2
Intuitive Technology Codes Assigned by Period and by Publication

Time Period/Publication	Number of Codes Assigned	Codes Assigned/Year
MISQ 1977–1989	6	0.5
MISQ 1990–1999	8	0.8
ISR 1990–1999	5	0.5
MISQ 2000–2009	8	0.8
ISR 2000–2009	4	0.4
MISQ 2010–2012	13	4.3
Total codes assigned	44	1.2

Source: McAran and Manwani (2013, 2014).

7.6   LIMITATIONS AND FUTURE RESEARCH

There are limitations to this research: (1) This research did not evaluate the effects of the TAII model over an extended period of time (longitudinally). (2) This research used self-reported use. (3) The research reached a segment of the legal profession target population who had sufficiently good skills to use internet social media, which was used to solicit respondents, and they may not be representative of the entire target population. (4) This research provided limited empirical data supporting the convergence of intuitiveness and ease of use. Further empirical research could be performed on the convergence of intuitiveness and ease of use. It would be particularly interesting to explore the differences in factors that influence user acceptance of technology among law, medicine, engineering, and accounting practitioners.

7.7   CONCLUSION

This research advances the research in the field of intuitive interaction in that it confirms the importance of intuitive interaction and its generalizability to other research domains and extends existing findings to technology acceptance, such that intuitive interaction with technology will have an increased likelihood of technology acceptance and use. The increasing convergence of the psychological constructs of intuitiveness and ease of use is identified, and support from the literature and additional evidence is provided concerning this emergent phenomenon. This convergence was integral to the results of the research and the consequent creation of the novel TAII model.

This research has identified PI as an emergent latent construct in technology acceptance. This was the first technology acceptance research that integrated a PI construct into TAM, and used legal technology as the basis of the research. This research has looked to the design of intuitive technology as a *design* solution to a long-standing problem in IS academic research: the issue of technology acceptance. Results from this research and from a review of the literature support the conjecture that there is an approaching convergence of the concepts of intuitiveness and ease of use. This research finding was unexpected and will require additional empirical research across differing technologies and technology platforms.

REFERENCES

Agarwal, R., & Prasad, J. (1997). The role of innovation characteristics and perceived voluntariness in the acceptance of information technologies. *Decision Sciences*, 28(3), 557–582.

Ajzen, I. (2012). The theory of planned behavior. In Lange, P. A. M., Kruglanski, A. W., & Higgins, E. T. (Eds.), *Handbook of Theories of Social Psychology*. London: Sage.

Apple. (2017a). Apple iOS human interface guidelines. Retrieved from https://developer.apple.com/ios/human-interface-guidelines/overview/design-principles/ (accessed March 18, 2017).

Apple. (2017b). macOS human interface guidelines. Retrieved from https://developer.apple.com/library/content/documentation/UserExperience/Conceptual/OSXHIGuidelines (accessed April 9, 2017).

Bakke, S. (2015). An activity theory approach to intuitiveness: From artefact to process. In *Proceedings of 17th International Conference on Human–Computer Interaction: Design and Evaluation (HCI International 2015)*, August 2–7, Los Angeles, 3–13.

Benbasat, I., & Barki, H. (2007). Quo vadis, TAM? Issues and reflections on technology acceptance research. *Journal of the Association for Information Systems*, 8(4), 211–218.

Blackler, A. (2006). Intuitive interaction with complex artefacts. PhD dissertation. Queensland University of Technology, Brisbane, Australia. Available from: https://eprints.qut.edu.au/16219.

Blackler, A., & Hurtienne, J. (2007). Towards a unified view of intuitive interaction: Definitions, models and tools across the world. *MMI-Interaktiv*, 13, 37–55.

Blackler, A., & Popovic, V. (2015). Editorial: Towards intuitive interaction theory. *Interacting with Computers*, 27(3), 203–209.

Blackler, A., Popovic, V., & Mahar, D. P. (2002). Intuitive use of products. In *Proceedings of Common Ground: Design Research Society International Conference 2002*, September 5–7, London.

Blackler, A., Popovic, V., & Mahar, D. P. (2010). Investigating users' intuitive interaction with complex artefacts. *Applied Ergonomics*, 41(1), 72–92.

Bödi, R. A., & Kaulich, T. W. (1992). Intuitive user interfaces (IUI): A CASE starting point for design and programming. *Computer Methods and Programs in Biomedicine*, 37(2), 69–74.

Burton-Jones, A., & Straub, D. W. (2006). Reconceptualising system usage: An approach and empirical test. *Information Systems Research*, 17(3), 228–246.

Chau, P. Y. K., & Hu, P. J. (2002). Examining a model of technology acceptance by individual professionals: An exploratory study. *Journal of Management Information Systems*, 18(4), 191–229.

Chuttur M. Y. (2009). Overview of the Technology Acceptance Model: Origins, developments and future directions. *Sprouts: Working Papers on Information Systems*, 9(37). Retrieved from http://aisel.aisnet.org/cgi/viewcontent.cgi?article=1289&context=sprouts_all (accessed June 25, 2015).

Davis, F. D. (1986). A technology acceptance model for empirically testing new end-user new information systems: Theory and results. PhD dissertation. MIT Sloan School of Management, Cambridge, MA.

Davis, F. D. (1989). Perceived usefulness, perceived ease of use, and user acceptance of information technology. *MIS Quarterly*, 13(3), 319–340.

Diefenbach, S., & Ullrich, D. (2015). An experience perspective on intuitive interaction: Central components and the special effect of domain transfer distance. *Interacting with Computers*, 27(3), 210–234.

Fantana, N. L., Riedel, T., Schlick, J., Ferber, S., Hupp, J., Miles, M., et al. (2013). IoT applications: Value creation for industry. In Vermesan, O., & Frie, P. (eds.), *Internet of Things: Converging Technologies for Smart Environments and Integrated Ecosystems*. Aalborg, Denmark: River, 153–206.

Fink, A. (2008). *Practicing Research: Discovering Evidence that Matters*. Thousand Oaks, CA: Sage.

Fishbein, M., & Ajzen, I. (1975). *Belief, Attitude, Intention and Behaviour: An Introduction to Theory and Research*. Reading, MA: Addison-Wesley. Retrieved from http://people.umass.edu/aizen/f&a1975.html (accessed May 8, 2013).

Fischer, S., Itoh, M., & Inagaki, T. (2015). Screening prototype features in terms of intuitive use: Design considerations and proof of concept. *Interacting with Computers*, 27(3), 256–270.

Fischer, S., Oelkers, B., Fierro, M., Itoh, M., & White, E. (2015). URU: A platform for prototyping and testing compatibility of multifunction interfaces with user knowledge schemata. In *Proceedings of the 17th International Conference on Human–Computer Interaction: Design and Evaluation (HCI International 2015)*, August 2–7, Los Angeles, pp. 151–160.

Glaser, B. G., and Straus, A. L. (2009). *The Discovery of Grounded Theory: Strategies for Qualitative Research* (4th paperback edition). London: AldineTransaction.

Gorinevsky, D. (2003). Lecture 5: Feedforward. Stanford University website. Retrieved from http://web.stanford.edu/class/archive/ee/ee392m/ee392m.1034/Lecture5_Feedfrwrd.pdf (accessed June 5, 2016).

Gregor, S. (2006). The nature of theory in information systems. *MIS Quarterly*, 30(3), 811–842.

Hair, J. F., Celsi, M., Money A. H., Samouel, P., & Page, M. (2016). *Essentials of Business Research Methods* (3rd ed). New York: Routledge.

Hair, J. F., Hult, G. T. M., Ringle, C. M., & Sarstedt, M. (2014). *A Primer on Partial Least Squares Structural Equation Modeling (PLS-SEM)*. Thousand Oaks, CA: Sage.

Hirschheim, R. (2007). Introduction to the special issue on "Quo vadis TAM? Issues and reflections on technology acceptance research." *Journal of the Association for Information Systems*, 8(4), Article 18.

Hevner, A. R., March, S. T., & Park, J., (2004). Design science in information systems research. *MIS Quarterly*, 28(3), 75–105.

Hurtienne, J. (2009). Image schemas and design for intuitive use: Exploring new guidance for user interface design. PhD dissertation. Technische Universität Berlin, Germany. Retrieved from http://opus.kobv.de/tuberlin/volltexte/2011/2970/pdf/hurtienne_joern.pdf (accessed May 5, 2015).

Hurtienne, J., & Blessing, L. (2007). Design for intuitive use: Testing image schema theory for user interface design. In *Proceedings of the 16th International Conference on Engineering Design (ICED'07)*, August 28–31, Cite des Sciences et de l'Industrie, Paris, France.

Islam, M. N. (2014). Design and evaluation of web interface signs to improve web usability a semiotic framework. PhD thesis, Åbo Akademi University, Department of Information Technologies, Turku, Finland.

Kappelgaard, A., & Bala, K. (2011). Comparison of intuitiveness, ease of use and preference in three growth hormone injection devices. In *Abstracts of the Pediatric Endocrinology Nursing Society Convention*, April 6-11 Indianapolis, IN.

King, W. R., & He, J. (2006). A meta-analysis of the Technology Acceptance Model. *Information & Management*, 43(6), 740–755.

Kovanović, V., Joksimović, S., Katerinopoulos, P., Michail, C., Siemens, G., & Gašević, D. (2017). Developing a MOOC experimentation platform: Insights from a user study, In *Proceedings of the Seventh International Conference on Learning Analytics and Knowledge (LAK '17)*, March 13–17, Vancouver, Canada, pp. 1–5.

Kroenung, J., & Eckhardt, A. (2015). The attitude cube: A three-dimensional model of situational factors in IS adoption and their impact on the attitude–behavior relationship. *Information & Management*, 52(60), 611–627.

Lakoff, G. (1987). *Women, Fire, and Dangerous Things*. Chicago, IL: University of Chicago Press.

Lambert, K. (1998). The relationship between computer interaction and individual user characteristics. MSc thesis, Rochester Institute of Technology, Rochester, New York. Retrieved from http://scholarworks.rit.edu/theses/8385 (accessed September 4, 2017).

Lohmann, S., Negru, S., & Bold, D. (2014). The ProtegeVOWL plugin: Ontology visualization for everyone. *ESWC 2014 Satellite Events, Anissaras, Crete, Greece, May 25–29, 2014: Revised Selected Papers*. Retrieved from http://www.springerprofessional.de/055---the-protegevowl-plugin253a-ontology-visualization-for-everyone/5382886.html (accessed December 21, 2014).

Macaranas, A., Antle, A. N., Riecke, B. E. (2015). What is intuitive interaction? Balancing users' performance and satisfaction with natural user interfaces. *Interacting with Computers*, 27(3), 357–370.

Marchal, D., Moerman, C., Casiez, G., & Roussel, N. (2013). Designing intuitive multi-touch 3D navigation techniques. In *Proceedings of 14th IFIP TCI3 Conference on Human–Computer Interaction (INTERACT'13)*, Cape Town, South Africa, pp. 19–36.

McAran, D. (2017). Integrating the intuitive into user acceptance of technology theory. DBA dissertation, Henley Business School, Henley-on-Thames, UK.

McAran, D., & Manwani, S. (2013). Characterizing the intuitive in information systems research. In *Proceedings of the 12th European Conference on Research Methodology for Business and Management*, July 4–5, University of Minho, Guimaraes, Portugal.

McAran, D., & Manwani, S. (2014). A gestalt generated from grounded theory concerning intuition in IS research. In *Proceedings of the 13th European Conference on Research Methodology for Business and Management Studies*, June 16–17, Cass Business School at City University, London.

McEwan, M., Blackler, A., Johnson D., & Wyeth, P. (2014). Natural mapping and intuitive interaction in videogames. In *Proceedings of the First ACM SIGCHI Annual Symposium on Computer–Human Interaction in Play (CHI PLAY'14)*, Toronto, Canada, pp. 191–200, October 19–22, 2014.

Moore, G. C., & Benbasat, I. (1991). Development of an instrument to measure the perceptions of adopting an information technology innovation. *Information Systems Research*, 2(3), 192–222.

Naumann, A., & Hurtienne, J. (2010). Benchmarks for intuitive interaction with mobile devices. In *Proceedings of the 12th International Conference on Human–Computer Interaction with Mobile Devices and Services (MobileHCI'10)*, September 7-10, pp. 401–402, Lisboa, Portugal.

O'Brien, M. A., Rogers, W. A., & Fisk, A. D. (2010). Developing an organizational model for intuitive design. Technical Report HFA-TR-100. Georgia Institute of Technology School of Psychology, Human Factors and Aging Laboratory, Atlanta, GA. Retrieved from http://smartech.gatech.edu/bitstream/handle/1853/40563/HFA-TR-1001 Intuitive DesignConceptualOverview.pdf?sequence=1 (accessed October 26, 2011).

Oxford Online Dictionary (2017). Feedforward. Oxford University Press. Retrieved from https://en.oxforddictionaries.com/definition/feedforward (accessed May 13, 2017).

Petter, S., DeLone, W., & McLean, E. R. (2013). Information system success: The quest for the independent variables. *Journal of Management Information Systems*, 29(4), 7–61.

Podsakoff, P. M., MacKenzie, S. B., Lee, J.-Y., & Podsakoff, N. P. (2003). Common method biases in behavioral research: A critical review of the literature and recommended remedies. *Journal of Applied Psychology*, 88(5), 879–903.

Qualtrics (2015). Retrieved from http://www.qualtrics.com (accessed May 28, 2015).

Raskin, J. (1994). Intuitive equals familiar. *Communications of the ACM*, 37(9), 17–19.

Ringle, C. M., Wende, S., & Becker, J. M. (2015). *SmartPLS 3*. Bönningstedt, Germany: SmartPLS.

Rogers, E. M. (1983). *Diffusion of Innovations (third edition)*. New York, NY: The Free Press.

Schmandt, C. (1980). Some applications of three-dimensional input. MSc dissertation. Massachusetts Institute of Technology, Cambridge, MA.

Shaw, N. (2011). Exploring systems usage at the feature level by reconceptualizing the dependent variable as a formative construct. DBA dissertation, Henley Business School, Henley-on-Thames, UK.

Sheppard, B., Hartwick, J., & Warshaw, P. (1988). The theory of reasoned action: A meta analysis of past research with recommendations for modifications and future research. *Journal of Consumer Research*, 15(4), 325–343.

Shirley, D. A., & Langan-Fox, J. (1996). Intuition: A review of the literature. *Psychological Reports*, 79, 363–384.

Straub, D. W., & Burton-Jones, A. (2007). Veni, vidi, vici: Breaking the TAM Logjam. *Journal of the Association for Information Systems*, 8(4), 223–229.

Straub, D., Limayem, M., & Karahanna-Evaristo, E. (1995). Measuring system usage: Implications for IS theory testing. *Management Science*, 41(8), 1328–1342.

Sutcliffe, A. (1988). *Human–Computer Interface Design*. London: Macmillan Education.

Taylor, S., & Todd, P. A. (1995). Understanding information technology usage: A test of competing models. *Information Systems Research*, 6(2), 144–176.

Tornatzky, L. G., & Klein, K. J. (1982). Innovation characteristics and innovation adoption implementation: A meta-analysis of findings. *IEEE Transactions on Engineering Management*, EM-29(1), 28–45.

Ullrich, D., & Diefenbach, S. (2010). INTUI: Exploring the facets of intuitive interaction. In Ziegler, J., & Schmidt, A. (Eds.), *Mensch und Computer 2010*, Interaktive Kulturen, Oldenbourg Publishing House, Munich, Germany: 251–260.

Van Ittersum, K., Rogers, W. A., Capar, M., Caine, K., O'Brien, M. A., Parsons, L. J. & Fisk, A. D. (2006). Understanding technology acceptance: Phase 1; Literature review and qualitative model development. Technical Report HFA-TR-0602. Georgia Institute of Technology School of Psychology, Human Factors and Aging Laboratory, Atlanta, GA. Retrieved from https://smartech.gatech.edu/bitstream/handle/1853/40576/HFA-TR-0705%20TechAcceptPhaseIII%20Quantitative%20Modeling.pdf (accessed July 15, 2017).

Venkatesh, V., & Bala, H. (2008). Technology Acceptance Model 3 and a research agenda on interventions. *Decision Sciences*, 39(2), 273–315.

Venkatesh, V., Davis, F. D., & Morris, M. G. (2007). Dead or alive? The development, trajectory and future of technology adoption research. *Journal of the Association for Information Systems*, 8(4), 267–286.

Venkatesh, V., Morris, M. G., Davis, G. B., & Davis, F. D. (2003). User acceptance of information technology: Toward a unified view. *MIS Quarterly*, 27(3), 425–478.

Vermesan, O., & Frie, P. (Eds.) (2013). *Internet of Things: Converging Technologies for Smart Environments and Integrated Ecosystem*. Aalborg, Denmark: River.

Wu, J., & Du, H. (2012). Toward a better understanding of behavioral intention and system usage constructs. *European Journal of Information Systems*, 21(6), 680–698.

8 Intuitive Interaction from an Experiential Perspective
The Intuitivity Illusion and Other Phenomena

Stefan Tretter, Sarah Diefenbach,
and Daniel Ullrich

CONTENTS

8.1 INTRODUCTION

Intuitive interaction as an experiential phenomenon consists of multiple facets—for example, effortless use or even the experience of magical interaction. What users experience and describe as "intuitive" may not necessarily parallel academic definitions of intuitive interaction. However, the subjective, experiential facets of product use can be crucial for product attractiveness and acceptance. This is also acknowledged in the general concept of *user experience*, particularly emphasizing the positive and emotional aspects of product use and the holistic experience of interacting with a system (e.g., Hassenzahl & Tractinsky, 2006). Thus, besides abstract definitions and criteria of intuitive interaction (e.g., the unconscious application of previously acquired knowledge; Blackler, Popovic, & Mahar, 2003; Naumann et al., 2007), it is also important to understand what forms an *impression* of intuitivity from the user's perspective, and in what respect this might differ from academic definitions or commonsense understanding of intuitivity.

In order to promote a phenomenological, experience-oriented perspective on intuitive interaction (see also Chapter 2; Blackler, Desai, McEwan, Popovic, & Diefenbach, 2018), we combine insights from user narratives and psychological theories of intuitive decision making comprising the INTUI framework. In the present chapter, we focus on one specific experiential phenomenon of intuitive interaction. In common sense, a product is judged as intuitive if one can handle it right away—without the need for initial training, tutorials, or reading the manual—since "the word intuitive insinuates that users can interact with a product using nothing more than their own intuition" (Monsees, 2014). However, our empirical research and daily life experience suggest that users also rate an interaction with an unknown product as intuitive if they were actually instructed how to use the product immediately beforehand. An attribution of the effective interaction to the instruction, which was actually crucial, is often missing. Instead, users judge the interaction as obvious and apparently disregard the instruction. In the course of this chapter, we will further elaborate on this, as we call it, *intuitivity illusion*.

Exploration of the intuitivity illusion thus provides an extended understanding of intuitive interaction and the related experiential phenomena. When users describe an interaction as intuitive, the application of prior knowledge could be truly unconscious and its origin actually inaccessible to the user (genuine intuitive interaction), or, alternatively, the source of knowledge could be theoretically accessible but not perceived as a crucial factor for successful interaction (intuitivity illusion). In the remainder of this chapter, we first summarize the central elements of and influencing factors on intuitive interaction as an experiential phenomenon (Section 8.2) and the particular role of retrospective judgments on (seemingly) intuitive interaction (Section 8.3). We then present empirical insights on the intuitivity illusion (Section 8.4) and discuss how these results can shape our theoretical understanding of intuitive interaction (Section 8.6.1) and how designers might utilize these findings (Section 8.6.2). We conclude with a remark on the intuitivity illusion as one example of the general relevance of conceiving intuitive interaction as an experiential phenomenon (Section 8.7).

8.2 INTUITIVE INTERACTION FROM AN EXPERIENTIAL PERSPECTIVE

Earlier research on intuitive interaction focused primarily on objective criteria, thereby relying on hard, quantifiable measures to define the concept and its inherent features (Blackler & Popovic, 2015). In recent years, however, there is growing interest in users' subjective experience when perceiving interactions as intuitive (see Chapters 2 and 4; Blackler et al., 2018; Blackler, Popovic, and Desai, 2018). To get closer to the subjective core of intuitive interaction, it seems reasonable to initially take a look at the underlying psychological term *intuition* incorporated in it.

Intuition is often talked about when people make decisions on a non-rational basis, following their affective experience of what feels right (Sadler-Smith & Sparrow, 2008). These judgments are based on an unconscious processing of prior knowledge, which is why the decision-making process is often hard to verbalize but feels rather effortless (Sinclair, 2011). Most interestingly, this goes hand in hand with a success rate that is higher than chance; that is, intuitively preferred choices usually turn out to be the right ones retrospectively (Sinclair, 2010). Products should be designed in a way that frames these kinds of decision processes; that is, when users act intuitively, the triggered behavior should result in a desired response (Still, Still, & Grgic, 2014). Although most recognized definitions of intuitive interaction acknowledge the aforementioned psychological processes—that is, the unconscious application of prior knowledge—the user's subjective experience of intuitive interaction cannot be measured by objective means (Blackler & Hurtienne, 2007). This emphasizes the importance of research on the subjective facets of intuitive use and how to assess them.

When we talk about interactive products today, their mere functionality is not the pivotal selling point anymore. Many products offer the same features, but it is those that provide a high degree of usability that stand out. According to international standards (Brooke, 1996), the usability of a product is defined by effectiveness (the ability of users to complete tasks using the system and the quality of the output of those tasks), efficiency (the level of resource consumed in performing tasks, generally measured as time), and satisfaction (users' subjective reactions to using the system). Effectiveness and efficiency are two components that correspond with the definitional objectives of intuitive use, while satisfaction can be understood as a subjective aspect that has not drawn the attention of earlier work.

A highly recommended approach to take product interaction one step further than usability is represented in the concept of user experience. It integrates traditional usability but also accounts for the beauty, hedonic, affective, or experiential aspects of technology use (Hassenzahl & Tractinsky, 2006). Ullrich and Diefenbach (2010) followed this phenomenological approach and focused on the subjective factors shaping the intuitive experience of users when interacting with products. Drawing on qualitative exploration and quantitative validation, their research resulted in the INTUI model, which postulates four subcomponents of intuitive product experience (Diefenbach & Ullrich, 2015). According to this study, product interactions can be evaluated on four facets, creating specific INTUI patterns that distinguish intuitive interaction experiences from one another. Each of these components poses another question regarding intuitive interaction design:

- *Effortlessness*: Can I act fast and without much effort?
- *Verbalizability*: Can I put into words how I proceeded and why?
- *Gut feeling*: Can I rely on my gut and what feels right?
- *Magical experience*: Does the interaction feel extraordinary, almost magical?

In addition, the model defines two levels of factors influencing the intuitive experience, as can be seen in Figure 8.1. The first level comprises the factors *product*,

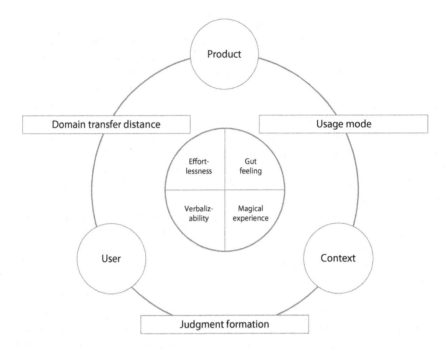

FIGURE 8.1 The INTUI model.

user, and *context* (Diefenbach & Ullrich, 2015). Apparently, the design of the *product* itself has an influence on the intuitive interaction experience. For instance, a simple change such as inverting an MP3 device's volume control reduces new user's ratings on its intuitivity (as explained by Still & Still [2018] in Chapter 3). In accordance with common definitions of intuitive interaction, *user* characteristics, foremost the relevant prior knowledge of a user, have a significant influence on the intuitive experience as well. Depending on how experienced a user is in a certain domain, the relevance of each INTUI component varyingly influences the intuitive experience (Ullrich & Diefenbach, 2010). Furthermore, it is obvious that the specific *context* of interaction is also substantial. When under time pressure, for example, it seems reasonable that the perceived effortlessness of an interaction is more weighted in a subsequent judgment about intuitivity than the perceived magical experience.

The second-level factors of the INTUI model can be understood as interplays of its first-level factors. *Usage mode* (*product × context*) is one of them. For instance, the role of verbalizability depends on whether interaction aims at a concrete goal (e.g., buying a concert ticket online) or is rather for the sake of exploration (e.g., browsing a news site; Hassenzahl & Ullrich, 2007). And even with the same product, verbalizability contributes more to the overall evaluation in goal modes (presumably because of a higher need for reflection and control) than in explorative modes (Ullrich & Diefenbach, 2011). The next second-level factor is *domain transfer distance* (*user × product*). It describes the distance between a domain of interaction and a corresponding domain that relevant prior knowledge stems from (see Chapter 2;

Blackler et al., 2018). With low transfer distance (e.g., two different generations of iPhones), verbalizability and effortlessness stand out, whereas interactions with a higher domain transfer distance (e.g., paper notebook and tablet) are characterized by gut feeling and magical experience. The second level is complemented by a third factor called *judgment formation* (*user* × *context*). In general, judgments about intuitive interaction are affected by a variety of impressions and the psychological processes of judgment formation. For example, even single non-intuitive functions can disproportionately impair the holistic rating of otherwise quite intuitive products, due to the effects of primacy or recency (Ullrich & Diefenbach, 2011). In the following section, we focus on one specific aspect of judgment formation—namely, retrospective judgments on (seemingly) intuitive interaction as a basis for the intuitivity illusion.

8.3 RETROSPECTIVE JUDGMENTS ON (SEEMINGLY) INTUITIVE INTERACTION

Intuitive interaction is considered a central factor for product success, especially in the field of consumer technology. People do not want to study manuals (e.g., Blackler, Gomez, Popovic, & Thompson, 2016), they just want to use the product *intuitively*. It is a widespread opinion that the Apple company's success is based on the fact that its products are designed with this fundamental principle in mind. The iPhone, especially, can be thought of as a pioneer product, and its features set milestones for future smartphone generations. Of course, that is not only on account of its functionality but because its interaction concept exploited the possibilities emerging from touchscreens in an apparently intuitive way.

One prime example of these intuitive interactions seems to be the standard zoom gesture—that is, the spreading of two fingers to enlarge the visual content displayed in between them. In fact, in a recent user survey ($N = 77$), most users described the zoom gesture as very intuitive; the average rating of intuitivity was 6.09 (SD = 1.22) on a seven-point scale (1 = not intuitive, 7 = very intuitive) (see Tretter, Ullrich, & Diefenbach, 2015). Also, users declared that they instantly knew how to perform the zoom gesture ($M = 6.08$; SD = 1.16)—a seemingly obvious case of intuitive interaction. However, when users were asked *how* they knew how to perform that gesture on first-time use, the range of answers was more diverse. A total of 52% answered that it was somehow intuitive or clear from the very beginning, 27% stated they had been introduced to it or had seen it depicted in advertisements, and the remaining 21% could not remember. Some participants made additional remarks emphasizing a combination of reasons, declaring they had seen the zoom gesture in advertisements; nevertheless, they also found it quite intuitive per se. Implicitly, this suggests that users assume they would have performed the zoom gesture spontaneously in the same way, even without the previous instruction through advertisements. This, however, could be a case of intuitivity illusion; in retrospect, the gesture seems intuitive and the only sensible way for a zoom function to work. However, we do not know whether users would have come to the same *a priori* judgment before the zoom gesture had been spread through advertisements and watching other iPhone users.

In sum, the example of the zoom gesture illustrates that very diverse paths can lead to the same positive judgment about a product's intuitivity, but not all of them represent what is understood as intuition in a classical sense. Though the majority of participants labeled the zoom gesture as very intuitive, only about half of them stated they had known how to execute it right away, while others reported instances of observational learning. Thus, it can be questioned whether the gesture is really what scholars define as intuitive.

The central question is whether the now-established zoom gesture was the most obvious option to naturally come to mind when wanting to enlarge an object, even before the iPhone era, or whether it is just quickly learnable and memorable? While this might seem a picky question, it has major theoretical and practical implications. Regarding research on intuitive interaction, such judgment errors could seriously impede the validity of user ratings. Allegedly representative users might not be able to correctly attribute their experience of intuitive interaction to observational learning, leading to false conclusions about the intuitivity of products. On a more practical level, companies might want to shift financial resources from development and design to broader and more intensive marketing campaigns, non-consciously instructing potential users about the correct usage (which was probably the case when the iPhone was introduced).

In our research, we found several examples of such an intuitivity illusion. In many cases, a product only appears intuitive if the successful interaction was actually enabled by previous instructions on how to interact with the product. However, people usually do not reflect on the original source of their knowledge. Subsequently, the intuitive impression is attributed to the product, even though it was not usable "right away" (as suggested by commonsense definitions of intuitive interaction) or from transferred existing knowledge (as per intuitive interaction scholars).

Indeed, there are corresponding interactions in the physical world that may help to understand the idea of the zoom gesture for enlarging a picture. For example, a drawing on a stretchable piece of material (e.g., a balloon) will magnify if you pull this material on two opposite ends, just like a digital image magnifies through the zoom gesture on a touchscreen. Or maybe users might have transposed the interaction concept of changing window size by dragging one of its corners with a mouse, known from various operation systems (although the iPhone lacked the cues that operation systems have in their task bar, and not only the whole screen can be stretched but also single parts). Still, the question remains whether people would have thought about this previous real-life interaction when in first contact with a smartphone device without prior instruction of any kind.

Supposedly, the iPhone's undo function serves as a more comprehensible example, as it is lesser known and probably not so tainted with prior experience. By shaking the device, intended by designers to be associated with discarding mistakes, the user prompts a request to undo the previous action. Even though this interaction concept might be inspired by other products such as Etch-a-Sketch (a formerly popular toy for simple drawings that could be erased by shaking), someone who might want to undo something would have quite a hard time finding out how to prompt the request. Yet, as soon as the user sees it, the appropriate association supports persistent learning at first glance. Similar to hidden-object pictures, those who know the solution can barely imagine how one cannot instantly see the allegedly obvious.

However, before empirical results backing the existence of the intuitivity illusion are presented, it is necessary to elaborate on the underlying, more general mechanisms of human perception and its potential flaws. The examples outlined here show that intuitive interaction as an experience cannot be solely dealt with using objective measures. Furthermore, the product is just one component shaping this experience, with context and user specifics contributing to it. When talking about intuitive interaction, it is mostly the user's prior knowledge that is focused on. But when intuitive interaction is approached phenomenologically as an experience and measured through users' statements, some characteristic errors arising from subjective judgments have to be considered. For instance, when rating the intuitivity of an interaction, impressions at the beginning and the end of an interaction are over-proportionally weighted (Ullrich, 2011). This primacy/recency effect is a typical example of cognitive biases in human perception (Mayo & Crockett, 1964). Most probably, it is not the only phenomenon that is playing a potential role in judgments about intuitive interaction. Its retrospective nature combined with accompanying unconscious processes make it prone to fallacy.

Numerous studies have shown so-called hindsight bias, according to which people in the aftermath of an event are prone to errors about their knowledge before the event occurred (Christensen-Szalanski & Willham, 1991). In particular, people overestimate the knowledge that they think they have after they are shown, for example, the solution to a knowledge task. In the context of product interaction, this means that it might be possible to, explicitly or implicitly, introduce users to specific knowledge necessary for interacting with a new product. The following successful interaction would elicit an intuitive experience that might be attributed, at least in part, to the product itself rather than to the instruction.

If that is the case, product interactions that people had no prior knowledge of could nevertheless feel intuitive given instruction *a priori*. Subsequently, the hindsight bias would make the product appear responsible for the experience and thereby induce some kind of intuitivity illusion. Still, a certain premise of this hypothesized effect is successful instruction. To this end, users have to be confronted with relevant knowledge and internalize it without an actual interaction. Hence, those interactions that are easily imitated seem especially prone to falling under an intuitivity illusion. On the other hand, if an interaction does not allow observational learning, successful usage is less probable and there is no intuitive experience, no biased attribution, and in the end no illusion.

With this in mind, some questions arise regarding the aforementioned iPhone feature. Many agree that the zoom gesture is intuitive—but why? Was it really intuition that guided people through the interaction? Or was it observational learning—that is, seeing someone else performing that gesture and its effects? Is the judgment about "intuitivity" actually hindsight bias? It is pointless to speculate now about how many new users would have intuitively shown the right behavior if it was not for the marketing. To clearly answer these kind of questions, an analogous situation has to be artificially created. Only if we know about the situation before someone observes an interaction for the first time can we tell how intuitive an interaction really is.

8.4 EXPLORING THE PHENOMENON OF THE INTUITIVITY ILLUSION

A straightforward way to empirically explore the intuitivity illusion is to compare the experiences of two groups and their retrospective judgments on interaction, one of them being instructed about the interaction concept beforehand, the other not. We can test whether the instructed group's attributions are biased and the intuitivity illusion really exists. If it shows that the instruction is, on the one hand, necessary to successful interaction (i.e., the group without instruction will not be able to perform the interaction intuitively), but on the other hand, the instructed group describes the interaction as intuitive, this means that attributions of high intuitivity may effectively rely on instructed interaction—that is, an illusion of intuitivity.

8.4.1 STUDY DESIGN

To this end, we conducted a computer experiment with 58 university students, where participants were to edit a sound file of spoken text in such a way that the single chunks fitted a specific stated order. To ensure that all participants conducted this task with the same prior knowledge, we introduced a novel product interaction—namely, specific mouse gestures. The question was how intuitive the participants would rate the interaction, depending on whether they received instruction or not. In order to copy, cut, or paste, participants had to press the right mouse button and perform one of three assigned movements to prompt the desired command. In three conditions, we altered the introduction presented before the task. The control group only received an introduction to the specific task ($N = 21$), while the other two groups were either assigned an additional manual ($N = 18$) or a video tutorial ($N = 19$) that instructed them about the mouse gesture interaction principles. Figure 8.2 provides an overview of the study's design.

Afterward, participants answered a questionnaire about the product and its use. At first, participants used the INTUI questionnaire to describe their interaction experience along 16 items, each allocated to one of the four components of

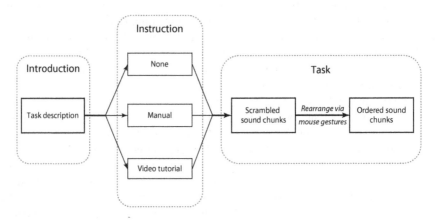

FIGURE 8.2 Overview of the study design.

the INTUI model, and one item assessing overall intuitivity (Diefenbach & Ullrich, 2015). These items are constructed with two opposite statements on each end of the response scale (e.g., "While using the product, I was guided by reason" vs. '... I was guided by feelings" for *gut feeling*). This was followed by seven-point scale items on overall product evaluation, the usefulness of the instruction, and the intuitivity of the mouse gesture interactions. Furthermore, participants who had been introduced to the mouse gesture interactions had to rate how likely someone would be to succeed at the task without any instruction. Additionally, participants had to indicate whether or not they faced difficulties when performing the task and answer an open question about what impeded or facilitated their performance. The questionnaire finished with items on potential compound variables (e.g., audio-editing proficiency) and demographics.

8.4.2 Results

Our study showed that previous instruction generally raised the probability of successfully performing the newly created interactions, which in turn led to higher effectiveness—that is, objectively better performances. In order to assess participants' task performance, three groups were defined and sound files were independently assessed by two raters. The three categories were labeled "0 = no recognizable changes to the original track," "1 = recognizable changes to the order of sound chunks," and "2 = all sound chunks placed in the right order." In fact, 85% of participants who were instructed via either video or manual fully completed the task (= 2), and 13% showed partial success (= 1). On the contrary, the control group without any previous instruction mostly ended up with no presentable achievement at all. Only 12% of the non-instructed participants showed at least partial success, and none were able to completely finish it. We conducted a Mann–Whitney U test to address the categorical nature of variables and the non-normal distribution of given data. Those participants who received instruction about the gesture interaction concept were significantly more effective at performing the tasks ($M_{instruction} = 1.81$, $SD_{instruction} = 0.46$; $M_{control} = 0.00$, $SD_{control} = 0.00$; $U = 10.50$, $p < .000$, $r = .90$). This pattern of results shows that the prior instruction was absolutely required for successful interaction. This forms an ideal basis to test for the intuitivity illusion, since the interaction can be claimed to be definitively not intuitive—at least, not understandable without prior instruction.

Additionally, among those who successfully performed the task, time to completion between the two conditions of instruction was analyzed. The group of participants who received video instruction did not perform better than those who were instructed via manual ($t[34] = 1.89$, $p = .068$, $d = .65$). There were also no significant differences between the two kinds of instruction regarding other measures, so in the following, data from these two groups are aggregated and reported as one collective instruction condition.

As the concept of the intuitivity illusion suggests, participants who successfully performed the interaction after prior instruction still judged the mouse gestures as intuitive. Overall, there was a positive Pearson product–moment correlation between effective task performance and overall judgment on intuitivity ($\rho = .54$, $p < .001$).

Regarding the specific mouse gesture interaction, those who had received previous instruction about the novel interaction reported significantly higher intuitive experiences ($M_{instruction}=4.73$, $SD_{instruction}=1.73$) than the control group ($M_{control}=1.62$, $SD_{control}=0.74$) based on a seven-point scale ($U=58.00$, $p<.000$, $r=.71$).

We also conducted a mediation analysis by means of the PROCESS macro (Hayes, 2012) to test for the assumed influence of instruction on intuitive experience through effective interaction. We chose this procedure since the involved bootstrapping resampling (1000 samples) accounts for deviation from normal distribution (Shrout & Bolger, 2002). Although the scales of measurement are not ideal due to their ordinal nature, the procedure can be considered robust in the face of these kinds of violations (West, Finch, & Curran, 1995; Norman, 2010). Mediation analysis confirmed the significant influence of the instruction condition on overall intuitive experience (total effect: $c=1.13$, $p<.000$, 95% CI [0.629; 1.631]) as well as on the assumed mediator task performance ($a=0.93$, $p<.000$, [0.761; 1.108]). If condition and performance are put into a joint model to predict intuitive experience, the influence of the mediator performance stays significant ($b=1.10$, $p=.004$, [0.376; 1.821]), while the effect of condition is partialed out (direct effect: $c'=0.10$, $p=.802$, [−0.718; 0.925]). The influence of condition on intuitive experience over the mediator therefore yields an indirect effect ($a*b=1.03$, $p=.004$, [0.211; 1.684]). The ratio of indirect effect ($a*b=1.03$) to total effect ($c=1.13$) shows that the mediator task performance explains 91% of the total effect. In sum, the effect of instruction on intuitive experience ratings was fully mediated by effective task performance.

When participants were asked how likely it was that a first-time user could manage the interaction without any pre-information, both groups rated this as rather unlikely (on a seven-point scale, 1=very unlikely, 7=very likely). The ratings of the instruction group ($M_{instruction}=2.81$, $SD_{instruction}=1.78$) and the control group ($M_{control}=2.05$, $SD_{control}=1.24$) did not differ significantly from one another.

Regarding the four INTUI components, there was no difference between the two groups who received previous instruction, but both significantly differed from the control group. Effortlessness ($M_{instruction}=4.90$, $SD_{instruction}=1.48$; $M_{control}=1.87$, $SD_{control}=0.86$; $U=47.00$, $p<.000$, $r=.73$), verbalizability ($M_{control}=6.25$, $SD_{instruction}=1.13$; $M_{control}=2.75$, $SD_{control}=1.55$; $U=33.00$, $p<.000$, $r=.77$), and magical experience ($M_{instruction}=4.04$, $SD_{instruction}=0.93$; $M_{control}=2.49$, $SD_{control}=1.11$; $U=110.00$, $p<.000$, $r=.59$) shaped the interaction experience of users who were instructed, while others mostly acted based on their gut feeling ($M_{instruction}=2.87$, $SD_{instruction}=1.17$; $M_{control}=4.69$, $SD_{control}=1.48$; $U=130.00$, $p<.000$, $r=.55$). Figure 8.3 illustrates the three groups' means on each of the four INTUI components and on their overall rating of intuitivity.

8.4.3 DISCUSSION

We conducted the study described here to explore the phenomenon of the intuitivity illusion and test for it in an experimental setting. Given the reported results, it becomes obvious that the objective criterion of task performance goes hand in hand with a more intense subjective feeling of intuitive interaction. As explained before, we assume that a cognitive bias occurs, since the instruction leads to an

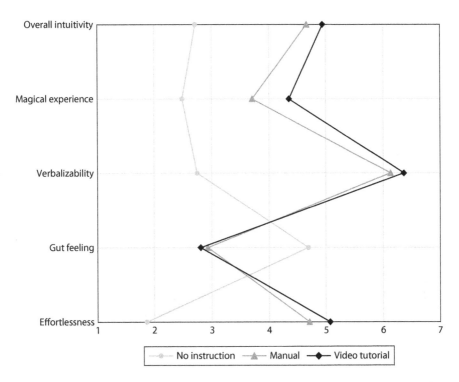

FIGURE 8.3 INTUI pattern for different instruction conditions.

effective interaction, but it is attributed to the intuitive usability of the product at hand. Empirical data thus support our assumption that an intuitivity illusion exists.

Interestingly, when participants were asked how likely it was that a first-time user could manage the interaction without any pre-information, participants in both conditions estimated this equally unlikely. Apparently, instructed users were well aware of the fact that they probably would not have been successful were they not instructed. It seems as though people are able to reflect on the given situation and come to a correct conclusion about their chance of success, although it obviously did not prevent them from describing the interaction as well as the product as intuitive. Accordingly, it can be stated that affective and cognitive judgments about the intuitivity of an interaction are to some degree independent from one another.

Open statements given by the instructed participants illustrate their tendency to attribute the successful interaction to the product's features, not the previous instruction. For example, the mouse gesture responsible for inserting previously cut elements was a movement similar to drawing a "V" on the interface. Several participants described this command as intuitive since "ctrl + V" is the widely known shortcut for pasting content from the clipboard on various computer operating systems, most notably Windows. "Intuitive," in this specific case, is rather a label for "easily memorable." Participants obviously tried to make sense of their experience in retrospect despite knowing that they would have probably failed to come to this conclusion if they had been left on their own—that is, had not been instructed on the interaction patterns. One person summarized it by stating, "I find it very intuitive, though someone without instruction would not have made it."

Additionally, it is also interesting to take a closer look at the four different INTUI components. As stated previously, the medium of instruction had no influence regarding the single components. However, after instruction, the interaction is shaped by a feeling of effortlessness and high verbalizability. This seems reasonable, since one only has to replicate the instruction encountered beforehand. Of course, these participants relied less on their gut feeling in turn, unlike non-instructed users. Without prior knowledge, however, following your gut is the only option available. Maybe the more magical experience of participants that received instruction can be attributed to the interaction concept in this specific study. Performing mouse gestures to draw different signs and initiate reactions from the program resembles in some way the schematic swinging of a wand.

Nevertheless, since the control group failed to successfully perform, an analysis of their intuitive interaction experience must be taken with caution. It will definitely be interesting to see future research comparing the INTUI patterns of two groups, with one being instructed and the other one consisting of some users who managed to perform the interaction without any introduction and some who did not. (Presumably, the instructed interaction would be characterized by a high degree of verbalizability and effortlessness, while spontaneous but successful interaction would be more characterized by gut feeling and magical experience.)

8.4.4   SUMMARY

Overall, our empirical exploration provides support for the existence of the intuitivity illusion as a phenomenon. Only the previous instruction enabled users to successfully interact with a product they would not have been able to interact with otherwise. However, users tended to construct a number of reasons for the intuitive characteristics of the product, even in cases where the instruction is presented in the most explicit way (i.e., by manual or video tutorial). Subsequently, when people are asked to reflect on the probability of success without any instruction, they are well aware of the substantial contribution of the previously presented instruction. However, other real-life examples (e.g., the zoom gesture discussed previously) suggest that this cognitive reflection will be reduced when a subtler kind of instruction (e.g., observational learning via advertisements or the vicarious experience of watching others) precedes the actual interaction.

8.5   RELATED PHENOMENA AND MECHANISMS

The present findings emphasize that to create an experience of intuitive interaction, it is helpful to design interactions in a way that allows users to retrospectively find a seemingly logical explanation of why an interaction is designed the way it is. In this way, even if a user cannot come up with the required interaction without any instruction, this will still make it easy to attribute the successful interaction to the product itself afterward. For example, this phenomenon might also be the reason for ratings on the intuitivity of the iPhone's zoom gesture or undo function. In retrospect, both gestures provide starting points for rational explanations as to why this specific gesture is the "only" appropriate and thinkable one. At first interaction, users might still have had the ability to reflect about their chances of successful interaction without prior learning and whether this gesture actually appears the most intuitive to them. However, over time, the attitude about its intuitivity was consolidated while the ability for correct attribution was lost.

Moreover, the intuitivity illusion shows similarities to a psychological phenomenon called the *illusion of explanatory depth*. Rozenblit & Keil (2002) showed in a series of studies that people often overestimate their understanding of the functioning of everyday artifacts and related physical mechanisms as well as a variety of other topics such as mental illnesses or economic markets. For example, their participants were very confident that they knew how a crossbow works until they were asked to show and explain it in more detail. After failing at the early stages, most of them became more aware of their real degree of expertise and adapted their statements significantly afterward. Similar to our study, where participants are misled by their subjective feeling of intuitivity, people fall prey to a naïve feeling about the depth of their own knowledge. But in both cases, they come to realize their delusion when asked to actively reflect on it. Both seem to enact comparable mechanisms, only differing in terms of to how much effort it takes to overcome the illusion from case to case. For example, in our experiment, participants dealt with an interaction completely new to them, making this easier. On the contrary, crossbows are well known, though not in great depth, obviously. Nevertheless, people have already established an attitude of familiarity toward them, promoting an illusion of knowledge about their function.

8.6 DEALING WITH INTUITIVITY ILLUSIONS: IMPLICATIONS FOR THEORY AND DESIGN

In some way, the present empirical findings and user judgments contradict the basic idea of intuitivity from a theoretical perspective. A product is not supposed to be intuitive if one has to be introduced to its usage. Once more, this emphasizes the relevance of considering intuitive interaction in a holistic sense, paying attention to all its facets and what it is in the end that creates an impression of intuitivity. Moreover, when dealing with intuitive interaction from a phenomenological perspective, the existence of such a phenomenon also offers opportunities and can be actively used and considered in design. The following sections discuss such theoretical and practical implications.

8.6.1 THEORETICAL IMPLICATIONS

We are talking about an intuitivity illusion in the sense that previous instruction substantially enables successful usage and thereby evokes an accompanying experience of intuitive interaction. Nonetheless, the term *illusion* depends on a notion of intuitive interaction that is described as the unconscious application of prior knowledge from a more or less distant transfer domain. In the case of the intuitivity illusion, this knowledge is obviously acquired within the same domain—that is, even the exact product. Thus, the appropriateness of the term *illusion* depends on researchers' concepts of intuitive interaction. In case of *inferential intuition* (Mohs et al., 2006), meaning an interaction that becomes intuitive over time through recurring practice, this would not be rated an illusion at all. In inferential intuition, the application of underlying knowledge becomes non-conscious over time but usually can be deliberately pulled back into consciousness. Under these conditions, when the interaction is accepted as a source of prior knowledge for itself—that is, the identical interaction—there would be no point in labeling it an illusion. With this understanding of intuitive interaction in mind, everything is at some point intuitive, given enough experience and exercise. Consequently, an instruction should be accepted as a legitimate source for conditioning intuitive interaction since it enables fast and successful usage of the product.

However, questioning whether something can be called an illusion comes down to where the theoretical border is drawn based on the definition at hand. With a phenomenological approach in mind, it seems reasonable to refer to subjective experiences instead of objective measures to answer the following questions: If an interaction can be fulfilled effortlessly, pre-consciously, and effectively because of repeated interactions in the past, would one really label this interaction intuitive? Or does there have to be at least some domain transfer distance? Would someone label an interaction intuitive if he or she knew it had to be performed multiple times previously? How much practice can an interaction demand and still be seen as intuitively designed?

Accordingly, future research, at least from a user experience perspective, should also consider the role of practice resulting from regular interactions over a longer period of time. The study results outlined here show that instructed participants reported an intuitive experience, but at the same time, the instruction had no effect on their estimation about whether a user would be able to interact without instruction. This suggests a differentiation between whether intuitivity is measured as a rating of someone's own intuitive interaction experience or someone's judgment about whether another person would be able to successfully interact without previous instruction. This would contrast the affectively influenced, experience-based judgment from a deliberate and rational general assessment. Following this approach, it can be assumed that only those interactions that draw on implicit knowledge and its non-conscious application would have an influence on both of these criteria. Since a user would not be aware of the relevant prior knowledge, he or she might infer another user's ability to successfully interact from their own. On the contrary, if knowledge and its source are known, intuitive interaction could nevertheless be experienced, but it should not affect the estimation of success for another, uninstructed user. The reported empirical results at least support this assumption. When participants were explicitly instructed, they gave high ratings on intuitivity but were still able to realistically estimate the chances of an uninstructed user's success.

However, this only applies to people's judgments immediately after the instructed interaction. If the zoom gesture is an example of the intuitivity illusion, why do everyday users not only experience intuitive interaction but also express with confidence that the interaction itself is intuitive? Time and repetition might serve as an explanation. As time goes by, people forget what enabled them to successfully interact with a product in the first place—instruction or intuitive design. With increasing distance in relation to the initial interaction, the specific learning situation fades out of memory and the product is not only affectively but also cognitively attributed with intuitivity. The former solely experience-based feeling spills over into a general judgment about intuitivity. Probably, it requires time as well as repetition to evoke this illusion. On the one hand, a long time after a single interaction with a product, one would not even implicitly remember the relevant prior knowledge for intuitive interaction. Thus, repetitions within that timeframe also matter. On the other hand, even with multiple repetitions, some time has to go by for the user to forget about the previous instruction.

Of course, this refers to a specific case where it was obviously hard to succeed without instruction. Nevertheless, there are two adaptions that can generally enhance the postulated processes. Firstly, even though there was (intentionally) no prior knowledge facilitating the acquisition of the mouse gesture interaction per se, the specific gesture-to-function mapping can be designed as variably intuitive—for example, assigning a slash gesture (/) instead of a circle gesture (○) to execute the "cut" command. This might enable users to more quickly internalize the new kind of interaction since it better resembles the real-life action of slicing. This supports a retrospective attribution to the product since it is easier for users to rationally make sense of the specific gesture afterward. Additionally, the more intuitive design fosters observational learning among users. Secondly, a subtler conveyance of knowledge could draw the users' attention away from the instruction and amplify the false attribution. Under a cover story, for example, participants could be presented with a visual advertisement showing the interaction—instead of an explicit video tutorial—to disguise that fact that they have just been instructed. In sum, the two key factors are well-designed single facets of the product ("I would have guessed it anyway" or "Oh, this makes sense") and a subtle method of instruction ("I got it all by myself"). On top of all these specific theoretical implications, the intuitivity illusion generally asks for a more sensible interpretation of user statements and a growing focus on instructions (or other sources of observational learning) when dealing with intuitive interaction.

8.6.2 IMPLICATIONS FOR DESIGN

Product developers should be particularly aware of the intuitivity illusion when it comes to assessing the intuitivity of a product by user ratings alone. On the other hand, they can also leverage this phenomenon by delivering relevant information on product interaction to the user—even if they are eminently conscious—who then tends to judge the product as more intuitive in retrospect. This provides manufacturers with the opportunity to establish a positive affect toward their products, even if there is a lack of product features that can be used right away. After all, a positive experience at first use is one of the main goals developers constantly aim for.

Given the effect of the intuitivity illusion, designers have to make up their minds about what they want to achieve. Intuitive interaction as well as "instructed" interaction have been shown to evoke a subjective positive impression, while effectiveness as an objective criterion is guaranteed too. Intuitive interaction allows users to engage in interaction right away and with low cognitive resources. However, there are plenty of situations when prior knowledge is not sufficient to fully exploit the functions of a product. At that point, when the possibilities of intuitive design reach their limits, instructions come in handy. Because of the intuitivity illusion, those instructions will not necessarily interfere with the subjective feeling of intuitive usability but can provide insights to the user that are sufficient for interaction—and serve the user with the desired positive experience.

It is good to know that due to attributional biases, it is not overly important how the user comes to the relevant knowledge in the first place but rather that he or she has such knowledge at all. Understanding the underlying principles of intuitive interaction can help designers to provide the necessary cues to trigger relevant prior knowledge. However, as soon as interactions become more complex, design alone does not always have the power to achieve this. This is where observational learning of any kind comes into play. The strongest effects probably occur when the cues are well-learned prior knowledge that the user is unaware that he or she had already been taught—that is, through marketing campaigns.

In reality, only a few commercial players have the chance to address potential future users through campaigns on a larger scale. According to present research, the most efficient way is to communicate all the relevant knowledge but in an implicit and unobtrusive way, so that corresponding instructions stay in the background while the product steps into the center of the attributional processes. Three premises for product design can be proposed. Firstly, basic interactions should follow conventional patterns—for example, a triangle for starting dynamic processes and a square for stopping them. Secondly, interactions that are probably not associated with prior knowledge have to be instructed—for example, holding the right mouse button to activate gesture tracking. And thirdly, instructed interactions, in turn, should again build on prior knowledge that enables fast observational knowledge—for example, performing a slash gesture to cut marked content.

Apart from these basic rules, the instructions themselves can be considered a "design element." While simple products allow for simple intuitive interactions, information technology products are often too complex to design all features in an intuitive way for all users. However, those products might allow the integration of instructions, which might make them even more suited to exploiting the intuitivity illusion (while also making undesired manuals obsolete). Information technology products can, for example, make use of the principles of gamification, engaging users in fun, challenging tasks, thereby implicitly instructing them on how to interact. *MessagEase* (Exideas, 2017) is an alternative keyboard application for smartphones where the nine most used letters of the alphabet are arranged in a 3×3 Sudoku-like grid. While these are selected by touching, the other 17 letters are placed around the nine and selected by swiping from the center of the key in the direction of the respective letter. Users learn about the functionality of this keyboard by starting with a simple game where balloons and parachutes float from the top to the bottom of the screen. These flying objects have to be destroyed by typing what is written on them. If they leave the display unharmed, the user loses virtual life points. As soon as users become somewhat acquainted, the difficulty increases through the addition of further letters up to whole words, thereby leading the learning process through unobtrusive, successive instruction.

Furthermore, the possibilities of information technology can be exploited in an even more sophisticated approach. On the one hand, there is common knowledge that can be expected for any user; on the other hand, users also differ in their specific knowledge. Hence, the gold standard are instructions that adapt to respective users' knowledge and interaction behavior. For example, information technologies could be designed in a way that users are successively presented with the knowledge they require in a certain situation. Adaptive instructions could react to users' current behavior, realizing what they are about to do and giving them successively more explicit hints on how to perform their desired next step. Thereby, the users are allowed to act based on their prior knowledge, giving them the freedom to explore as long as they are successful with their interactions, only interfering when help is needed. For instance, EA Sport's recent FIFA football video games incorporate a corresponding feature.

> If you press R3 during play, the game brings up a "contextual overlay system" above the player you are controlling, providing a selection of possible moves. If you're just dribbling the ball in your own half, it'll probably just suggest short or long passes, together with the relevant button, but once in the opponent's territory it'll bring up a range of attacking options. There are several levels of advice to select from, so intermediate and experienced players can get more complex pointers, including tougher skill moves.
>
> **(Stuart, 2015)**

As can be seen, information technology is ideally situated to leverage attributional biases, since the instruction and the product are no longer two entities but one. Thus, another important strand for future research on intuitive interaction seems to be a deeper exploration of *integrated instructions* and the question of how to deliver information about interaction concepts in a smooth way. All this can add to the user's impression of intuitivity and a fulfilling user experience.

All these are opportunities for designers to make use of the intuitivity illusion and promote an intuitive experience with the help of intelligent instructions. And finally, over time, even memories of initial difficulties in the interaction with the product might vanish and leave the user convinced about how intuitive their product is. A perfect mesh of design and psychological bias will promote a positive user experience and create happy customers.

8.7   CONCLUSION

In conclusion, the intuitivity illusion is just one demonstration of the general relevance of conceiving intuitive interaction as an experiential phenomenon. From a user experience perspective, it is about creating the basis for a fulfilling, positive experience of interaction, and the impression of intuitive interaction is one such building block. Metaphors, image schemas, and other techniques (Blackler & Hurtienne, 2007; Hurtienne & Israel, 2007; Antle, Corness, & Droumeva, 2009) provide a helpful approach to integrating previous existing knowledge into interaction concepts and to foster intuitive interaction. But interaction built on newly acquired knowledge might also be experienced as intuitive, depending on how that new knowledge is communicated. As discussed in the preceding paragraphs, there are various opportunities to communicate interaction concepts to the user, which might be experienced as more or less explicit forms of knowledge communication. In effect, successful interaction might be attributed to a greater or lesser extent to instruction—or simply intuition.

REFERENCES

Antle, A. N., Corness, G., & Droumeva, M. (2009). Human–computer-intuition? Exploring the cognitive basis for intuition in embodied interaction. *International Journal of Arts and Technology*, *2*(3), 235–254.

Blackler, A., Desai, S., McEwan, M., Popovic, V., & Diefenbach, S. (2018). Perspectives on the nature of intuitive interaction. In A. Blackler (Ed.), *Intuitive Interaction: Research and Application* (pp. 19–40). Boca Raton, FL: CRC Press.

Blackler, A. L., & Hurtienne, J. (2007). Towards a unified view of intuitive interaction: Definitions, models and tools across the world. *MMI-interaktiv*, *13*, 36–54.

Blackler, A., & Popovic, V. (2015). Towards intuitive interaction theory. *Interacting with Computers*, *27*(3), 203–209.

Blackler, A., Popovic, V., & Desai, S. (2018). Intuitive interaction research methods. In A. Blackler (Ed.), *Intuitive Interaction: Research and Application* (pp. 65–88). Boca Raton, FL: CRC Press.

Blackler, A. L., Gomez, R., Popovic, V., & Thompson, M. H. (2016). Life is too short to RTFM: How users relate to documentation and excess features in consumer products. *Interacting with Computers*, *28*(1), 27–46.

Blackler, A., Popovic, V., & Mahar, D. (2002). Intuitive use of products. In D. Durling & J. Shackleton (Ed.), *Common Ground: Proceedings of the Design Research Society International Conference 2002*, September 5–7, London.

Blackler, A., Popovic, V., & Mahar, D. (2003). The nature of intuitive use of products: An experimental approach. *Design Studies*, *24*(6), 491–506.

Brooke, J. (1996). SUS: A quick and dirty usability scale. *Usability Evaluation in Industry*, *189*(194), 4–7.

Christensen-Szalanski, J. J., & Willham, C. F. (1991). The hindsight bias: A meta-analysis. *Organizational Behavior and Human Decision Processes*, *48*(1), 147–168.

Diefenbach, S., & Ullrich, D. (2015). An experience perspective on intuitive interaction: Central components and the special effect of domain transfer distance. *Interacting with Computers*, *27*(3), 210–234.

Exideas.com (2017). *MessagEase*. Retrieved from http://www.exideas.com/ME/index.php.

Hassenzahl, M., & Tractinsky, N. (2006). User experience: A research agenda. *Behaviour & Information Technology*, *25*(2), 91–97.

Hassenzahl, M., & Ullrich, D. (2007). To do or not to do: Differences in user experience and retrospective judgments depending on the presence or absence of instrumental goals. *Interacting with Computers*, *19*(4), 429–437.

Hayes, A. F. (2012). *PROCESS: A Versatile Computational Tool for Observed Variable Mediation, Moderation, and Conditional Process Modeling*. New York: Guilford Press.

Hurtienne, J., & Israel, J. H. (2007). Image schemas and their metaphorical extensions: Intuitive patterns for tangible interaction. In *Proceedings of the TEI'07, 1st International Conference on Tangible and Embedded Interaction* (pp. 127–134). ACM, New York, NY.

Mayo, C. W., & Crockett, W. H. (1964). Cognitive complexity and primacy–recency effects in impression formation. *The Journal of Abnormal and Social Psychology*, *68*(3), 335–338.

Mohs, C., Hurtienne, J., Israel, J. H., Naumann, A., Kindsmüller, M. C., Meyer, H. A., & Pohlmeyer, A. (2006). IUUI: Intuitive use of user interfaces. In T. Bosenick, M. Hassenzahl, M. Müller-Prove, & M. Peissner (Eds.), *Usability Professionals 2006* (pp. 130–133). Stuttgart, Germany: Usability Professionals' Association.

Monsees, J. (2014). Putting "intuitive" back into intuitive design. *UX Magazine*, October 14. Retrieved from https://uxmag.com/articles/putting-intuitive-back-into-intuitive-design.

Naumann, A., Hurtienne, J., Israel, J. H., Mohs, C., Kindsmüller, M. C., Meyer, H. A., & Hußlein, S. (2007). Intuitive use of user interfaces: Defining a vague concept. In D. Harris (Ed.), *Engineering Psychology and Cognitive Ergonomics* (pp. 128–136). Heidelberg, Germany: Springer.

Norman, G. (2010). Likert scales, levels of measurement and the "laws" of statistics. *Advances in Health Sciences Education*, *15*(5), 625–632.

Rozenblit, L., & Keil, F. (2002). The misunderstood limits of folk science: An illusion of explanatory depth. *Cognitive Science*, *26*(5), 521–562.

Sadler-Smith, E., & Sparrow, P. R. (2008). Intuition in organizational decision making. In G. P. Hodgkinson & W. H. Starbuck (Eds.), *The Oxford Handbook of Organizational Decision Making* (pp. 305–324). Oxford: Oxford University Press.

Shrout, P. E., & Bolger, N. (2002). Mediation in experimental and nonexperimental studies: New procedures and recommendations. *Psychological Methods*, *7*(4), 422–445.

Sinclair, M. (2010). Misconceptions about intuition. *Psychological Inquiry*, *21*(4), 378–386.

Sinclair, M. (2011). *Handbook of Intuition Research*. Cheltenham, UK: Edward Elgar.

Still, J. D., & Still, M. L. (2018). Cognitively describing intuitive interactions. In A. Blackler (Ed.), *Intuitive Interaction: Research and Application* (pp. 41–62). Boca Raton, FL: CRC Press.

Still, J. D., Still, M. L., & Grgic, J. (2014). Designing intuitive interactions: Exploring performance and reflection measures. *Interacting with Computers*, *27*(3), 271–286.

Stuart, K. (2015). FIFA 16: The eight key new features. *Guardian*, Jun 23. Retrieved from https://www.theguardian.com/technology/2015/jun/23/fifa-16-eight-key-new-features.

Tretter, S., Ullrich, D., & Diefenbach, S. (2015). Die Intuitiväts-Illusion: Intuitives Nutzererleben durch Attributionsfehler. In S. Diefenbach, M. Pielot, & N. Henze (Eds.). *Mensch und Computer 2015: Tagungsband* (pp. 283–286). Munich: De Gruyter.

Ullrich, D. (2011). Primacy- und Recency-Effekt bei der Produktinteraktion und ihr Einfluss auf die Bewertung von Inuitivität und User Experience. In *Proceedings 7. Tagung der Fachgruppe Arbeits-, Organisations- und Wirtschaftspsychologie* (pp. 124–125). Rostock, Germany: Universität Rostock.

Ullrich, D., & Diefenbach, S. (2010). INTUI: Exploring the facets of intuitive interaction. In J. Ziegler & A. Schmidt (Eds.), *Mensch und Computer 2010* (pp. 251–260). Munich, Germany: Oldenburg.

Ullrich, D., & Diefenbach, S. (2011). Erlebnis intuitive Interaktion: Ein phänomenlogischer Ansatz. *i-com Zeitschrift für interaktive und kooperative Medien*, *10*(3), 63–68.

West, S. G., Finch, J. F., & Curran, P. J. (1995). Structural equation models with non-normal variables. In R. Hoyle (Ed.), *Structural Equation Modeling: Concepts, Issues, and Applications* (pp. 56–75). Thousand Oaks, CA: Sage.

Part III

Applying Intuitive Interaction

9 City Context, Digital Content, and the Design of Intuitive Urban Interfaces

Luke Hespanhol

CONTENTS

9.1   INTRODUCTION

The drive toward urbanization has become one of the defining trends of the twenty-first century. Cities are already the de facto habitat of humankind, with recent data showing that more than 54% of humanity lives in urban environments (United Nations, 2014). The majority of us spend most of our lives navigating our way across relatively high-density environments as loose members of a crowd often little connected by anything but the urban precincts we share. At the same time, pervasive computing, ubiquitous internet access, and increasingly responsive environments have turned modern city living into a highly mediated experience enacted through a continuous interplay between physical and digital interfaces. While expanding the possibilities for data access and social interactions in public spaces, this permanent mediation of city life and the proliferation of digital interfaces also pose significant challenges to interaction designers, architects, and urban planners: how can we more seamlessly interact with the increasingly digitized environment around us, and to what extent does such an environment facilitate or hinder interactions between us?

To a large extent, seamless and intuitive interactions with the built environment are subject to the same sort of tension between nature and nurture experienced by other interactive artifacts. On the one hand, *physical affordances* (Chapter 2; Blackler, Desai, McEwan, Popovic, & Diefenbach, 2018) leverage the physical properties of our bodies—including position in space relative to other elements in the environment, scale, and the perceptive apparatus provided by our senses—to favor certain types of interaction over others. On the other hand, as pointed out by Blackler and Hurtienne (2007), *population stereotypes* and cultural standards play a large role in the ways we as humans appropriate, make sense of, and utilize interactive artifacts, lending meaning to their affordances to fulfill our preconceived expectations. Yet, culture does not evolve in a vacuum; the production and reproduction of global "culture" is therefore largely forged by—while also feeding back into—the very day-to-day interactions and face-to-face relations locally experienced in cities (Hannerz, 1996; Kokot, 2007), illustrating what Appadurai (1991) has referred to as *global cultural flows*.

In this chapter, I propose to investigate this feedback loop by analyzing the influence of city contexts and digital contents in the design of intuitive urban interfaces. I start by outlining related research into urban interfaces (de Waal, 2014; Tomitsch, 2016) and the model proposed by Hespanhol and Tomitsch (2015), demonstrating how intuitiveness can be designed into public spaces by the thoughtful placement of technology in relation to the layout of the urban area and the choice of media modality for feedback to passers-by. I then attempt an expansion of that model by proposing additional core elements and analyzing their various arrangements from the perspective of well-known combinations of affordances, metaphors, and population stereotypes (Chapter 2; Blackler et al., 2018) for the utilization of the public space, arriving into a taxonomy of patterns—or urban interfaces—for intuitive interaction with and within the urban built environment. Following that, I employ the concepts of plug-in architecture and interfaces (Hespanhol & Tomitsch, 2018) to support the analysis of three distinct case studies through the lenses of the identified patterns. I conclude by discussing the implications of the analysis for the under-standing of intuitive urban interfaces, compiled as a series of design guidelines and recommendations.

9.2 RELATED WORK

Research about the use of digital technologies in public spaces has followed a simi-lar pattern as that observed for technology applied to other domains, particularly in the sense that it is often focused on making sense of past experimental applications, analyzing the approaches taken, and deriving heuristics and insights that can then inform future design efforts. Particularly important, in that regard, are the views derived from the human–computer interaction of interpreting digital media in the built environment as yet another interface and the analysis of intuitiveness in public spaces derived as a consequence.

9.2.1 DIGITAL MEDIA AND THE BUILT ENVIRONMENT

A significant volume of recent research has focused on the increasing adoption of digital media in public spaces, its role in the design of urban precincts, and its impact on social dynamics and public behavior. De Waal (2014) looked at the city through the lens of interaction design and cultural studies, comparing it to a soft-ware interface, and enquiring on the functions of public spaces, the experiences citizens derive from them, and how digital media and technologies can assist—or hinder—the experience and usability of cities. Tomitsch (2016) took the analogy further by comparing the urban built environment to a computer operating system and proposing the concept of *city apps*, which could be deployed to, run on, or be removed from the *smart city* just like a mobile app on a smartphone. City apps would therefore correspond to lightweight pieces of information and functional-ity, running on situated urban interfaces and dedicated to the provision of specific services or experiences to passers-by in those locations. Examples could include projection-based interactive information consoles running on the window of a building (Tomitsch, Ackad, Dawson, Hespanhol, & Kay, 2014), impromptu commu-nity polling interfaces rolled out to existing urban screens (Hespanhol et al., 2015; Schroeter, Foth, & Satchell, 2012), or even media façades (Behrens, Valkanova, Fatah gen Schieck, & Brumby, 2014).

As with previous kinds of interactive interfaces, of course, these new develop-ments also beg the question of how to design for intuitiveness—that is, how to create urban interfaces with clear, sensible affordances that people can easily make sense of and promptly interact with.

9.2.2 INTUITIVENESS IN PUBLIC SPACES

Urban interaction is still to a large extent an emerging field, and, as pointed out by Blackler and Hurtienne (2007), "for very new technology which has none of its own conventions, a metaphor which relates to something that is familiar to the users would need to be applied" (p. 7). In that regard, Blackler, Popovic, and Mahar (2010) stress the importance of exposure to the same features across different contexts in order for those features to become perceived as intuitive. Technology familiarity therefore emerges from their analysis as a significant factor determining the level of intuitiveness of an interface, summarized into three basic principles of intuitive interaction: (1) the use of familiar features from the same domain, (2) the transfer of familiar things from other domains, and (3) redundancy and internal consistency (Blackler, Mahar, & Popovic, 2010).

One way, therefore, to identify design solutions perceived as intuitive is to analyze recent case studies in regards to particular design variables, and by doing so identify emerging recurrent patterns in the values adopted for those. In that vein, Hespanhol and Dalsgaard (2015) have conducted an analysis of media architecture implementations since the turn of the twenty-first century and compiled a list of seven patterns for the design of social interaction in public spaces. Their study made a distinction between two broad categories of design factors, which the authors use to analyze case studies and derive the design patterns: (1) *layout factors*, such as the spatial layout of the urban precinct and types of feedback given to citizens, and (2) *social interaction modes*, reflecting more experiential qualities of interactive urban environments, such as self-expression, playfulness, and collective narratives. These factors highlight the continuous interplay between *carriers* (physical structures, including the architecture and layout of the built environment as well as lighting and hardware infrastructure) and *content* (the nature of the actual concepts articulated through the digital and electronic media), and their roles as key contributors to the quality of the interactive experience. They also point to the relevance of the embodied experience in public spaces and the impact of both the nature of the digital media and their placement into the built environment, insofar as the media prompt physical and perceived affordances for users of the public space.

Antle, Corness, and Droumeva (2009) have demonstrated, however, that mental models derived from physical affordances are not by themselves sufficient to imply intuitiveness. Drawing from the work of Lakoff and Johnson (1980) on metaphors and embodied schemata, they investigated intuitiveness in responsive environments based specifically on full-body interaction. Embodied schemata are mental representations of recurring dynamic patterns of bodily interactions that structure the way we understand the world; embodied metaphors conceptually extend embodied schemata through the linking of a source domain that is an embodied schema and a target domain that is an abstract concept (e.g., the body's general upright position in space implies various spatial metaphors based on a vertical hierarchy). Results from their studies demonstrate that a mapping constructed on embodied metaphors leads indeed to greater non-conscious interaction; however, this appears not to be sufficient: in order to be readily and quickly learned, the mapping should also be easily discoverable.

In the context of public spaces, and in light of the analysis by Hespanhol and Dalsgaard (2015) discussed previously, this discoverability can be facilitated by (1) feedback modalities, (2) the type and placement of interactive technology (particularly regarding how to gather input from the public), and (3) the layout of urban precincts.

9.3 CONSIDERATIONS FOR THE DESIGN OF INTUITIVE URBAN INTERFACES

In this section, I discuss these three aspects identified previously in light of the earlier model for intuitive urban interfaces proposed by Hespanhol and Tomitsch (2015) and identify potential aspects for revision and expansion of such a model by articulating some of the ideas emerging from recent literature in the field.

Hespanhol and Tomitsch (2015) presented an analysis of interactive design strategies for intuitive interaction, using as examples urban interventions developed during their research as well as those developed by others. These case studies were then analyzed through the lenses of the conceptual frameworks for intuitive interaction developed by both Blackler, Popovic, and Mahar (2006) and the Intuitive Use of User Interfaces (IUUI) research group (Blackler & Hurtienne, 2007; Hurtienne & Blessing, 2007; Hurtienne & Israel, 2007), pointing out factors that support the intuitiveness of each scenario. Notably, they identified the nature of the feedback provided to people engaging in interaction in public spaces as being particularly relevant for intuitiveness. This can be chiefly articulated in three ways: (1) the *specificity* of the feedback provided—that is, whether it is directed to specific individuals or scattered across the environment; (2) the *timing* of the feedback—that is, if it is immediate and concrete or delayed and abstract; and (3) the type of feedback *modality*, mostly distinguishing between visual and audio feedback (Hespanhol & Tomitsch, 2015).

Just as important as the type of feedback offered is the level of accessibility of the responsive public space, how those two factors interplay, and how different combinations promote certain types of social behavior over others. The earlier model by Hespanhol and Tomitsch (2015) proposed a classification of interactive public spaces into three types: *performative, allotted,* and *responsive ambient*. Performative interfaces are those with well-delimited interactive zones, often supported by local signage or other visual elements making clear where they are and small enough to accommodate only a small number of participants. That, as a consequence, leads to a natural division of the public into *performers* (active participants) and *spectators* (passive participants), effectively transforming the interactive zone into a stage. The second type of interactive public space, allotted, shares the same basic features as performative ones; however, they are larger precincts, enough to accommodate a small crowd, so that each participant can interact with the responsive environment within their own "lot." As a consequence, a different set of social dynamics are set in motion, with the interaction spread across larger sections of the urban precinct and participants prompted to negotiate the space spontaneously among themselves. The third category, responsive ambient, refers to urban interfaces that track and react to the presence and movements of passers-by—however, without addressing them directly: the feedback offered is mostly indirect and abstract, rather than responding to specific individuals.

In a later publication, Hespanhol and Dalsgaard (2015) complemented that initial classification by also taking into account the spatial layout of public spaces. Based on the concepts of *spatial nodes* and *links*, as described in urban planning works such as Hillier and Hanson's *The Social Logic of Space* (1984), they divided public spaces, respectively, into two broader categories: (1) plazas and (2) thoroughfares. They defined a plaza as a wide, open public space where a large number of citizens potentially congregate, facilitating social encounters as well as passive social practices such as people watching or even loitering. A thoroughfare, by contrast, is a transit area connecting plazas, therefore characterized by the continuous flux of passers-by walking from one destination to another. As Hespanhol and Dalsgaard (2015) observed in their analysis of case studies from the past 15 years, the number of plazas largely outweighs thoroughfares in the design of interactive urban media architecture, reflecting the tendency to place responsive interfaces in areas of the city that already play the role of destinations in the mental model of citizens, rather than the spaces that connect them.

9.3.1 INPUT MECHANISMS

Another aspect of the analysis by Hespanhol and Dalsgaard (2015) that had not been included in the model by Hespanhol and Tomitsch (2015) is the nature of the input mechanisms used for the interaction, in addition to the feedback strategies.

Designs of large interactive public environments have traditionally resorted to three main strategies for input mechanisms: (1) mobile devices, (2) consoles, and (3) gesture-based full-body interactions. The first leverages on the increasing ubiquity of handheld, portable, and computationally powerful devices (particularly smartphones and tablets) as a mediator between users of a public space and specific digitally augmented spots (usually electronic displays or media facades) within the precinct. Those devices offer the advantage of minimizing a participant's public exposure and are consequently relatively effective in encouraging opportunistic interaction (Kray, Galani, & Rohs, 2008) by reducing the perceived risk of social embarrassment. On the flip side, the utilization of mobile devices for such a purpose is novel and unrelated to their usual mode of use, compromising the intuitiveness of the interface. This poses challenges to the scalability of the interface, raising issues regarding concurrent access between parties that are largely unaware of each other. Moreover, it also poses a practical barrier to the spontaneous emergence of interaction in the urban space, since it generally requires people to install an app or navigate to a website, which in turn needs to be communicated.

The second strategy, and a slightly more inclusive way to articulate interaction in public spaces, is to use a mediator between the person interacting and the surrounding digital environment, in the form of a tangible user interface (Behrens et al., 2014) such as a kiosk or console. Consoles are generally in prominent positions within a public space but manage to counterbalance such a public profile by allowing a somewhat private interactive experience.

The third strategy for input mechanisms, gesture-based interfaces controlled via full-body interaction, has generally come to be associated with the term *natural user interfaces* (NUIs), which would imply intuitiveness. NUI refers to settings that allow interaction with digital systems to take place without the presence of any specific device operated or worn by the user, leveraging from people's learned skills of using their bodies to interact with the world. Yet, it can be argued that many of the in-air gestures proposed as innate are distant from the actions people would intuitively take when dealing with most products or digital interfaces. For example, despite many of the so-called NUIs being designed for touchless mid-air gestures, not all of them are highly intuitive, as pointed out by Chattopadhyay and Bolchini (2015), who compared two touchless gesture primitives: one consisting of 3D strokes, requiring previous expertise about how accurately to execute them, and the other consisting of 2D directional strokes, which drew on sensorimotor abilities. As the researchers found, making accurate 3D strokes is less intuitive for the general population than making 2D strokes, precisely for requiring prior exposure to (and thus knowledge of) the primitives. In other words, although such tracking technologies present a necessary step toward more natural interactive mechanisms (in the sense of being based solely on the human body), in their current state, they can hardly be regarded as natural, or intuitive, themselves.

9.4 AN EXPANDED MODEL FOR INTUITIVE URBAN INTERFACES

Based on the previous discussion, I propose to extend the model for the design of intuitive interactions in public spaces originally posed by Hespanhol and Tomitsch (2015) by combining it with some of the aspects later employed by Hespanhol and Daslgaard (2015) in their analysis of social interaction design patterns for urban media architecture—namely, public space layouts and input mechanisms. In support of that extension, I argue that the utilization of the most common input mechanisms for public interaction (mobile devices, consoles, and full-body interaction) is framed by familiarity derived by their use in different domains, which, as pointed out by Blackler and Hurtienne (2007) and explained in more depth in Chapter 2 (Blackler et al., 2018), is one of the core factors leading to intuitive interaction. Likewise, the layouts of different public spaces enact embodied metaphors and mental models of movement through the public space, which, as demonstrated by Antle et al. (2009), lead to the non-conscious adoption of well-established patterns of interaction with the environment and others within it.

9.4.1 Public Space Affordances and Urban Interaction Archetypes

Leveraging on the literature on intuitive interaction, one can to a large extent attribute the emergence of those patterns of interaction to what I hereby define as *urban interaction archetypes*: combinations of affordances, metaphors, and population stereotypes (Chapter 2; Blackler et al., 2018) for the utilization of the public space and tacit social norms for interactions corresponding to the urban context. Urban interaction archetypes thus constitute expressions of conventions for navigation and the use of physical and digital elements of public spaces understood by whole populations, and leveraging sensorimotor and cultural knowledge established through experiences of particular urban situations. The different combinations of public space types and feedback strategies proposed by Hespanhol and Tomitsch (2015) with the public space layouts and input mechanisms adopted by Hespanhol and Dalsgaard (2015) result in a range of public space affordances—here defined as the cumulative set of physical and perceived (Norman, 2013) affordances of the various elements constituting a publicly accessible and shared area within the city. Examples of physical public space affordances include sitting on a bench or public staircase while reading a book, leaning against bridge railings, or throwing litter in garbage bins. Perceived public space affordances are seen as equivalent to familiar features (Blackler, 2008), either previously encountered in cities (e.g., a pedestrian pressing a button to turn traffic lights green for them and red for cars) or borrowed from other domains (e.g., moving your body to vote on civic issues (Hespanhol et al., 2015), an interactive feature inspired both by shadow play and video games using depth-view cameras). Because public space affordances can be traced back to existing cultural conventions, they can be used to lay out a taxonomy of urban interaction archetypes to guide the design of intuitive interactions in public spaces.

9.4.2 CONTEXTUAL VARIABLES AND URBAN INTERFACES

For the purposes of this analysis, I define as contextual variables the four dimensions involved in the analysis of Hespanhol and Tomitsch (2015) and Hespanhol and Dalsgaard (2015): namely, (1) type of interface or interactive public space, (2) layout of the public space, (3) input mechanism, and (4) feedback strategy. Figure 9.1 shows the values for each variable considered for this analysis.

Following that, I define an urban interface as a socio-technical city context consisting of a particular arrangement of those variables, enabling a particular city context in the public space. For example, a performative interaction deployed to a plaza, enabled through mobile device–based interaction with visual feedback, is a very different urban interface from a thoroughfare that responds with ambient audio to full-body presence.

9.4.3 DERIVING URBAN INTERACTION ARCHETYPES FROM URBAN INTERFACES: THE URBIA MODEL

By laying out the different combinations of the contextual variables defined previously, we can uncover a range of urban interfaces. Furthermore, based on the nature of the interactions prompted by those interfaces, it is possible to establish a mapping between each of them and the corresponding urban interaction archetype that would underpin intuitive interaction in each context. For integrating the design aspects from the model originally proposed by Hespanhol and Tomitsch (2015) with the preceding

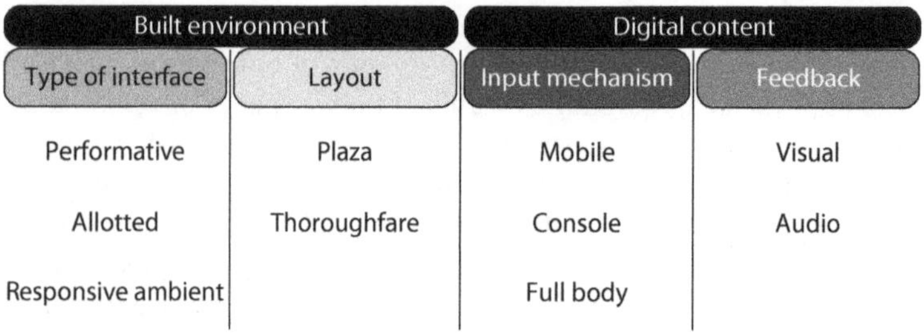

Built environment		Digital content	
Type of interface	Layout	Input mechanism	Feedback
Performative	Plaza	Mobile	Visual
Allotted	Thoroughfare	Console	Audio
Responsive ambient		Full body	

FIGURE 9.1 Contextual variables.

contextual variables, derived in turn from Hespanhol and Dalsgaard (2015), such a mapping constitutes an expanded model for intuitive urban interfaces, which I hereby refer to as the Urban interaction archetypes Model, or Urbia Model. Figure 9.2 illustrates the Urbia Model and its range of 18 urban interaction archetypes mapped to each combination of contextual variables (shown as white rectangles in Figure 9.2 and in italics throughout the chapter).

Each box in Figure 9.2 represents a design specification defined by a particular combination of the contextual variables, and thus defining a "script" for interaction in the resulting urban interface or, as defined in Section 9.4.1, a particular urban interaction archetype. In that regard, they are not dissimilar to design patterns, a common analytical exercise where designers look back at a range of case studies and attempt to identify recurrent forms of organizing design variables to produce specific outcomes. Introduced by Alexander (1977) in his studies on architectural practice and history, the notion of design patterns was later adopted by many others, including Gamma, Helm, Johnson, & Vlissides (1995) in software engineering and Hespanhol and Dalsgaard (2015) in media architecture. Just like denominations given to design patterns (e.g. *building complex* or *wings of light* by Alexander [1977]; *bridge*, *flyweight*, or *chain of responsibility* by Gamma et al. [1995]; and *amusement park* or *smooth operator* by Hespanhol and Dalsgaard [2015]), the labels in Figure 9.2 are somewhat arbitrary and chosen for illustrative purposes. Yet, as with design patterns, the underlying specifications they represent stand as interaction archetypes for the type and scope of interactions afforded by them. In the following sections, I describe each of the identified interaction archetypes in respect to their corresponding contextual variables.

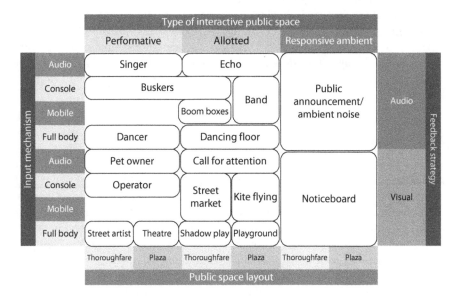

FIGURE 9.2   The Urbia Model for intuitive urban interfaces.

9.4.4   RESPONSIVE AMBIENT INTERFACES

Responsive ambient interfaces are those that, despite responding to inputs from the audience, are purposely designed not to appear interactive (Hespanhol & Tomitsch, 2015). There may be many reasons for this: for example, the work may be installed in a zone where the congregation of a large crowd is undesirable. Regardless of the input or sensing mechanism actually adopted, the audio and/or visual feedback is designed so as not to appear to be responding directly to explicit interaction by individuals in the space; rather, the design tries to spread their attention to elements in the environment around them. Feedback is therefore more diffused, often delayed, and with the passer-by only realizing its occurrence after the event, in an expression of the landing effect (Mueller, Walter, Bailly, Nischt, & Alt, 2012). When the feedback is purely audio based, it takes the general form of *ambient noise*; when content is structured to communicate meaning directly to the group of people in the precinct, it can also take the form of a *public announcement*. Crucially, it addresses the whole audience, regardless of who actually triggered the response. Likewise, the spatial layout of the precinct bears no influence on the way the responsiveness is structured, since passers-by do not actually stop or organize themselves in space in order to engage in interaction.

When visual feedback is used, the ambient nature of the environment takes the archetype of a visualization of cumulative data gathered from multiple passers-by over time or a visual effect in reaction to actions performed by people in the precinct. Again, no visual feedback is directed at specific individuals, with the interface assuming the archetype of a *noticeboard*. By that, I refer to situations when visual elements in a public space change as a result of input from people, yet not in an immediate manner as a direct interaction with a device or display would do; rather, the changes happen over a period of time. While still a two-way interface, the visual feedback resulting from an individual's action in the public space is not necessarily displayed immediately after the person has performed it. *Chromapollination*, presented by Hespanhol and Tomitsch (2015), is a successful example of a responsive environment making use of delayed visual feedback (in this case, LED lights) to create a dynamic responsive environment without inviting passers-by to linger in the space.

9.4.5   ALLOTTED INTERFACES

Interfaces that are large enough to accommodate multiple participants engaging in interaction are defined as allotted (Hespanhol & Tomitsch, 2015) if they assign each participant a section (or lot) of the broader interface, in which they can enact a meaningful interactive experience. In terms of their positioning in space, participants can be either stationary or free to move around; in the former scenario, the interaction usually happens through kiosks or consoles installed in the space; in the latter, people can interact with their mobile phones, voices, or by moving their bodies in front of the sensor (full-body interaction). Also, when participants are moving, negotiation of space and ownership of sections of the interface can naturally emerge, giving rise to a swarm pattern (Hespanhol & Tomitsch, 2015), with communication between participants unfolding both directly through physical interaction in the space and via triangulation through the digital interface.

It is possible to identify a few different interaction archetypes for allotted urban interfaces. For example, given the pervasive nature of audio, direct feedback to individuals becomes challenging. The effects of audio feedback thus vary considerably depending on the input mechanism adopted: scattered audio input results on an experience akin to *echo*, with immediate feedback sound mapped to the sound produced by an individual in space and somewhat directed at them. In a plaza-like spatial configuration—where people tend to linger in the space, perceived in turn as a destination—participants playing with physical instruments (consoles or mobile devices) and producing sound back to the environment enact the archetype of a *band* performing in the space. In thoroughfares, however, where other people would most likely be walking by the space, the mobility of the device leads to different city contexts: players interacting with instruments fixed in the space (consoles) assume the role of *buskers*, while those playing mobile instruments producing sound are like people carrying portable *boom boxes*. Finally, if a group of people engage in full-body interaction in public space, producing audio feedback in turn, the urban context becomes akin to a *dance floor*.

Visual feedback alters the contexts significantly, as it can more effectively direct replies to specific individuals. Visual feedback in response to audio input by scattered individuals resembles a collective *call for attention*, as if participants were trying to "wake the system up." For the remaining input strategies, the type of spatial layout—and therefore the rate of movement by the multiple participants in the space—can produce different kinds of urban interface archetypes. For instance, in a thoroughfare context, people interacting with instruments (consoles or mobile devices) and producing visual feedback create a context resembling a *street market*, through which "performances" other members of the public walk, and occasionally stop for further appreciation. A plaza layout, however, creates a context where multiple participants congregate in a single space and watch, from their positions, their projected identities interfere with each other, visually, in the broader, shared interface—like people in a park *flying kites* and constantly negotiating their flight paths. An example of such a context is the iRiS smartphone app (Boring et al., 2011) developed for concurrent interaction by multiple participants, who can simultaneously "paint"—by selecting colors from a palette in their phones—the shared media facade of the Ars Electronica Center in Linz, Austria, changing the colors of its panels accordingly. Participants were not always capable of seeing each other but shared the attention of the interface that the visual feedback was given from and thus communicated to each other via triangulation. For full-body interaction, when enacted in a thoroughfare, direct visual feedback resembles *shadow play*, with the digital content directly mimicking participants. In plazas, however, people interacting with their bodies on their allotted parts of the interface assume a configuration akin to a *playground*, as in the *Solstice LAMP* light installation presented by Hespanhol et al. (2014).

9.4.6 Performative Interfaces

Unlike allotted interfaces, interaction in performative public spaces (Hespanhol & Tomitsch, 2015) is restricted to a small number of participants, often to only a single individual. Any interaction thus results in an immediate division of the public on-site: the performer (i.e., the person engaging in the interaction) and everyone else, who become the performer's impromptu audience. In such a context, audio input by a performer resulting in audio feedback characterizes a performance similar to the one observed with *singers* playing in public events. Someone playing an instrument onstage producing solely audio functions as a *busker*. Full-body performers reacting to sound behave like *dancers*.

As with allotted interfaces, visual feedback creates different urban interfaces and dynamics in performative interfaces. Interaction via audio resulting in visual feedback works like participants effectively attempting to "tame" and "control" the interface through verbal communication, often repeating commands until the interface complies with its orders, much like a *pet owner* to their animal. One example of such a context is one of my previous public installations, *LOL* (Hespanhol, 2016), which I will also discuss in Section 9.5.5. Interaction via an instrument (console) producing visual feedback in a performative setting results in a context akin to an *operator* performing with equipment in front of an audience—that is, somebody who is clearly in charge of an interface that is in turn not affected by anyone else but its operator. One example of such a setting is the *Smart Citizen Sentiment Dashboard* by Behrens et al. (2014), where a console operated by public transport smartcard technology was placed in a public space in front a large-scale media façade; passers-by could then use their personal cards to express their feelings about different civic topics by tapping them into specific sections of the console, which was then translated into corresponding graphical visualizations in the media façade. As the console only allowed interaction by one card at a time, it was clear who was operating the media façade at any given time. Also, it is fair to say that such a context emerges regardless of the spatial layout of the space. Full-body interaction, however, unfolds differently with visual feedback in plazas or thoroughfares: in the latter, the performer appears to passers-by as a *street artist* working with their body to produce a visual expression and is mostly subject to fleeting encounters with a transient audience; in the former, however, performers expresses themselves in front of a largely captive audience, like an actor on a *theater* stage. *Aarhus by Light* by Brynskov et al. (2009) or *Body Movies* by Lozano-Hemmer (2001) are well-known case studies of performative, full-body interaction with visual feedback.

9.4.7 Non-applicable User Interfaces

As it is possible to see from Figure 9.2, the combinations involving mobile devices in performative public spaces were left blank. The reason for that has to do with the definition of performative public spaces put forward by Hespanhol and Tomitsch (2015): a public interface capable of accommodating only one or a few people interacting, who thus become performers in the space. A mobile device would only fit that description insofar as its users remained close to the same spot while interacting, so that an audience can develop around them. However, in that case, they would be indistinguishable from consoles—hence the omission of mobile devices in performative public spaces and the inclusion of consoles.

9.5 DISCUSSION

The preceding model proposes urban interaction archetypes for each of the combinations of contextual variables that define the different types of urban interfaces. Conversely, it suggests that when a particular combination is in place, the corresponding archetype defines the interactive context most likely to be intuitively assumed by members of the general public. It poses, therefore, conditions for the design of particular forms of interaction, as well as helping to explain situations where people misinterpreted urban interfaces, as I will discuss through the following case studies.

9.5.1 CITY CONTEXTS: PLUG-IN ARCHITECTURE AND PLUG-IN INTERFACES

One of the benefits of mapping urban interaction archetypes to combinations of design variables, as in the model presented previously, is that it becomes clearer what are the predominant types of interaction that may emerge in a public space given a certain spatial layout, the nature of the interface, the availability of certain input mechanisms, and the ways feedback is provided. In this way, the model helps to define the scope of most likely interactions for a given public space, based on how intuitively the interfaces will be approached and interacted with by people based on pre-established archetypes of public space utilization. That, in turn, supports the deployment of *plug-in architecture* (Hespanhol & Tomitsch, 2018), which is defined as the set of design properties observed in or assigned to a built environment that enables the accommodation of *plug-in interfaces*. Those, in turn, are defined as portable, interactive media technology, deployed directly to public spaces on a temporary basis and leveraging existing urban infrastructure and social dynamics. Consequently, plug-in interfaces lead to specific *plug-in choreographies* as a set of new social dynamics unfolding in the public space as a consequence of the deployment of a plug-in interface. The archetypes defined in Figure 9.2 correspond, therefore, to various plug-in choreographies supported by each of the plug-in interfaces resulting from the different combinations of the four contextual variables presented in this chapter (Figure 9.1): type of interface, layout of the public space, type of input mechanism, and type of feedback provided.

9.5.2 EXAMPLES OF INTUITIVE INTERACTION IN URBAN INTERFACES

What the concept of plug-in architecture also means, however, is that an inadvertent combination of contextual variables can lead to appropriation of the interface by passers-by in ways that have not been anticipated by the designer. While that can lead to novel forms of interaction, it also potentially compromises the intended use of the space—for example, people assuming one archetype when the designers intended another. In the following sections, I will present case studies that illustrate different occurrences of such mismatches.

In addition to the kind of input mechanism and feedback strategy employed, the nature of the actual digital content provided to people in the public space can also play a significant role in making an urban interface more or less intuitive. When investigating the possibilities of plug-in architecture, Hespanhol and

Tomitsch (2018), for example, have conducted a series of studies on polling inter-faces deployed to public spaces, with two basic variations of the interface, and two variations of the deployment site. The first interface consisted of a survey running as a web application on a tablet; the second consisted of an audio-based device sensing the proximity of passers-by and asking them binary (i.e., "Yes/No") questions via audio when they came close enough to the sensor. People would then respond to the question by placing their hands over one of two different light sensors, one labeled "Yes" and the other labeled "No." The two chosen locations were both thoroughfares: the first, a pedestrian crossing—where people normally have to pause for a couple of minutes—and the second, a pathway through a public park. The researchers also varied the placement of the devices—on a pole or on a stand—thus resulting in eight different configurations. Results of comparisons between interfaces strongly endorsed the effectiveness of technology familiarity in promoting participation when mapping content to the input mechanism of the interfaces: all three setups with the highest rates of conversion of passive passers-by into active participants in the interaction employed the iPad interface. Across all setups, the iPad interface also resulted in a conversion rate more than 3.5 times higher than the one produced by the audio interface. Despite the lack of explicit signage guiding the interaction, and the fact that neither interface constitutes by itself a familiar feature in the context of public spaces, passers-by still managed to swiftly make sense of the tablet-based setups and engage in interaction with that interface much more successfully than with the audio device. This is consistent with the literature on intuitive interaction, which argues that an interface will be perceived as intuitive if used in similar contexts to those it is normally found in or, if used in a different context, it follows the same interactive rules as those in its original context (Blackler & Hurtienne, 2007). Indeed, a web survey running on a tablet works the same way in public as it would be expected to run in any other context. As a consequence, passers-by could make sense of it immediately, unlike the audio device.

Another related example is provided by Steinberger, Foth, & Alt (2014), who developed *Vote with Your Feet*, a tangible plug-in interface exploring the social and spatial affordances offered by a bus shelter. The interaction mechanism was very straightforward and intuitive: a digital screen, mounted at the roof of the bus shelter, would display "Yes/No" questions, one at a time. Once a question was dis-played, people could cast their votes by stepping on one of two tangible buttons on the ground, one labeled with "Yes," the other with "No." In the context of the Urbia Model (Figure 9.2), it provided a plaza (i.e., a destination) configuration for a console-mediated performative interaction (one participant at a time, on a well-marked spot), guided by visual feedback. It therefore successfully enacted the *operator* archetype.

Naturally, the design of urban interfaces does not always succeed in being so unambiguous and intuitive. The proposed Urbia Model offers an expanded array of lenses for analyzing the intuitiveness of existing case studies, as well as understand-ing their shortcomings. In the following sections, I will analyze three urban inter-faces that prompted intuitive appropriations by members of the public in ways that did not correspond to the intentions of their designers.

9.5.3 APPROPRIATION OF A RESPONSIVE AMBIENT INTERFACE

Silent Island was an audiovisual installation I conceived in 2016 for Vivid Sydney, an annual festival of light, music, and ideas in Sydney, Australia. It consisted of a shipping container surrounded by a timber roof and wall structure serving a dual purpose: (1) to provide a single-access path to and from the interior of the container, with clearly marked entrance and exit points; and (2) to provide sufficient acoustic insulation in the interior of the container from the exterior environment, so that an individual standing inside would hardly be able to hear noise produced outside. Acoustic insulation was further enhanced by cladding the interior of the shipping container with thick architectural foam, which also worked as a canvas for the projection of digital graphics to the interior walls of the enclosed environment. Microphones were placed on the top of the structure, outside, and the external environment noise was then translated into animated projections inside. The intention was to create an environment subverting ordinary sensorial experience, where visitors could find refuge from the external noise pollution of the city, while still being able to appreciate the enhanced perception of its shifting audio quality and intensity through its expression via visual aesthetics.

Importantly, the installation was deliberately designed to be inconspicuous from the outside, only having lighting to indicate entrance and exit doors, in order to avoid giving out false affordances to visitors. For the same reason, it was designed as a responsive ambient environment, without providing consoles for interaction or reacting directly to any full-body movement or audio produced by visitors.

Yet, the appropriation of the environment by people could often be observed. For example, the continuously high numbers of visitors turned the environment into a thoroughfare most of the time, with people constantly flowing through the space and only spending a few minutes inside. During that short period of time, two interesting behaviors were recurrently observed: (1) Visitors would use their bodies to position themselves in front of the projectors and cast shadows into the interiors walls of the space; over time, that evolved into increasingly creative and collective shadow playing, as illustrated in Figure 9.3. (2) Some visitors noticed that, if they jumped inside the space, they could produce new interesting patterns in the projected visual graphics, and so they kept doing so, which prompted imitation by other visitors. The reason for that effect was that jumping made noise and also caused the whole structure to shake, including the microphones. The resulting noise was then immediately visualized in the interior wall projections, closing the interaction feedback loop.

When analyzing the interaction appropriation observed from the perspective of the urban interaction archetypes presented in Figure 9.2, it appears clear that an issue with *Silent Island*'s design was that, despite being intended to work as an ambient visual *noticeboard* for the exterior noise pollution, it actually also allowed other archetypes to manifest themselves. The thoroughfare's spatial layout, combined with the possibility of visitors to use their full body for *shadow play*, meant that the space was at times interpreted as allotted. That impression was then further crystallized by the immediate visual feedback derived every time someone stomped their feet into the floor, enacting the *call for attention* archetype and repeatedly attempting to make the environment respond to their input.

FIGURE 9.3 Emerging interaction through the appropriation of a non-interactive interface.

9.5.4 Appropriation of an Allotted Interface

The Climate on the Wall (Dalsgaard & Halskov, 2010) consisted of a large projection on a wall of the Ridehuset, a prominent historical building in a busy intersection in the center of Aarhus, Denmark, conceived to be in operation during the climate conference Beyond Kyoto, taking place in the city in 2009. The projection showed falling speech bubbles containing words commonly used in the ongoing discussions about climate. Cameras installed along the façade were used to detect people passing by. As passers-by walked along the projections, they could then "grab" a word, carry it along, and reposition it further down, thus having the ability to contribute to the climate debate by forming phrases expressing their views. The interface was perceived as playful, and passers-by would occasionally stop and interact with the projections. More often than not, however, implicit or inadvertent interaction (Mueller et al., 2012) would take place: people would unwillingly engage in interaction without noticing, with some only later realizing they had been dragging a speech bubble along with them. The lack of proper grammar seemed to hinder the formation of coherent sentences, but, most importantly, the best spot to observe the evolution of graphics on the screen was from the opposite side of the street—that is, by people watching the facade, not those actually interacting with it. The space—designed as allotted—failed to even be perceived as interactive since, first, people would inadvertently input their position, and second, they were not able to properly see and make sense of the feedback given to them by the projected digital content. Considering the model presented in Figure 9.2, the space was designed so that passers-by would intuitively understand the interaction mechanism proposed, based on the use of full-body interaction for enacting *shadow play*. However, given that participants could not effectively make sense of their own shadows, the environment ended up functioning as a responsive ambient visual *noticeboard*, reflecting the scrambled words dragged by people as they walked by the wall.

9.5.5 Appropriation of a Performative Interface

The final example of appropriation of the interactive properties of an urban responsive environment by passers-by is *LOL*, an interactive light installation that I presented at Vivid Sydney in 2014 (Hespanhol, 2016). The installation consisted of 10 inflatable balls resting on a platform, floating on the waters of an urban marina. Each ball would display a projection of an animated human mouth, which in the absence of external input would remain shut. When noise above a certain threshold was detected in the nearby environment—via a microphone concealed in one of the local benches—the mouths would burst into loud laughter, enticing the public to laugh back and, consequently, keep the interaction loop going. Conversely, once noise fell under the set threshold, the projections would return to rest.

Admittedly, a public projection that responds to laughter—or noise, for that matter—is not a very common and familiar occurrence, therefore posing challenges for designing affordances able to convey existing urban interaction archetypes. In an attempt to make the interface more intuitive, I assigned different noise thresholds for each projected mouth, so that varying levels of input audio would make the installation respond accordingly. Occasional noise would cause only a couple of mouths to react; sustained, loud noise would make them all respond. The design solution was thus intended to be straightforward in enabling performance: input noise could be given from anywhere along the water's edge in front of the floating balls, watched by anyone standing in the surroundings, and all feedback was given by the installation from the balls in the water. The only physical element installed on land had no digital content and consisted of a waterproof equipment box installed on the pedestrian deck overlooking the water and housing the projector, speakers, and a camera.

That, however, seems to have been sufficient to lead to a lack of clarity about the ways the interaction worked, and to appropriation by the public attempting to make sense of an unfamiliar urban interface. In observations of 900 passers-by, about 41% (374/900) attempted to interact with the installation. From those interactions, about 8% (30/374) of participants attempted to trigger the laughter on the projections by trying to tamper with the equipment box. The remaining 92% (344/374) either attempted to make noise on either side of the box, where no physical affordance was present (337/374), or tried to tamper with the box after seeing someone else doing so (7/374). This was, after all, a plaza context, where the interaction had been designed as performative through the use of audio to create a feedback loop of reciprocal audio and visuals. Yet, the presence of the equipment box as the only existing tangible element on land, next to the projections, led many people to interpret the urban interface through the *operator* archetype, as if the box might actually function as a *kiosk* or *console* through which the projections could be controlled.

These examples demonstrate how predictive of intuitive interaction—or not, as the case may be—are the urban interaction archetypes listed in the Urbia Model, and which archetypes people will intuitively expect to find for each city context. *Silent Island*, for example, was designed as a *noticeboard* in a responsive ambient environment; however, the manner in which the audio input and visual feedback were employed prompted people to make sense of it through the *shadow play* and *call for attention* archetypes. *The Climate on the Wall* led to the opposite effect: it was designed as an allotted interface where interaction would unfold though the *shadow play* archetype; however, the failure to enable passers-by to perceive the effects of their shadows on the interface encouraged its perception as a mere visual *noticeboard*. Finally, the unintended perceived affordances of the physical setup for *LOL* prompted people to interpret the interaction as following the *operator* archetype, when the installation had in fact been designed as a performative/allotted space that people would interact with by making noise (the *echo, call for attention,* and *pet owner* archetypes).

9.6 IMPLICATIONS FOR DESIGN

The preceding discussion raises two very important considerations for the design of intuitive urban interfaces: (1) the impact of false archetypes and (2) the impact of context and content.

9.6.1 IMPACT OF FALSE ARCHETYPES

The Urbia Model presented in Figure 9.2 maps various combinations of values for four relevant contextual variables—thereby defining different urban interfaces—to archetypes describing the interactions that could intuitively emerge from their deployment in public spaces. The flipside of that argument, as illustrated by the preceding examples, is that the inadvertent deployment of any of those combinations would also result in those very same dynamics unfolding, intuitively from the perspective of the people interacting, but possibly contrary to the original plan set by the designer. People will, after all, try to make sense of an unfamiliar environment by falling back on the recollection of previous similar situations they may have experienced. In the absence of visual feedback unambiguously related to the interaction, other visual elements in the immediately surrounding area become perceived affordances, even if they are only part of the local infrastructure. In the case of *LOL* (Hespanhol, 2016), the somewhat imposing presence of the equipment box right in front of the projections might explain why many people interpreted it as a console to trigger the laughter from the projected mouths floating in the bay. In *Silent Island*, the change in the patterns of projections resulting from the accidental shaking of the installation structure was enough to suggest to visitors that jumping around in front of the installation was an effective way to impact their surroundings.

As described by Blackler, Popovic, and Mahar (2010), technological familiarity plays a very large role in determining the intuitiveness of an interface: as humans, when trying to make sense of new technological settings, we seek potential clues in experiences we had with previous interfaces. In the context where *LOL* was set, the equipment box was the only visible element on the deck close enough to the projections to be perceived as an affordance (Norman, 2004). Until people became aware of the audio-interactive nature of the work, it made good sense to regard the equipment box as some sort of console or kiosk. That is especially true when one considers recent similar implementations using tangible user interfaces as mediators between media facades and the general public (Behrens et al., 2014; Fischer & Hornecker, 2012). As indicated by Hespanhol and Tomitsch (2015), when faced with an interactive public environment, people have become familiar with receiving direct feedback to their actions, usually in the form of mirrored images that can convey agency and identity. As a consequence, when becoming aware of the presence of an installation in public spaces, people initially try to make sense of it by moving their bodies, gesturing to it, or touching parts of the interface; if any of those attempts yields any perceived feedback—as in the case of *Silent Island*—the exploration of the urban interface hardly continues any further, thus compromising the whole experience intended by the designer. Conversely, in the absence of visual feedback matching their individual input, the environment may be intuitively perceived either as ambient (i.e., faintly responsive or not interactive at all)—as was the case with *The Climate on the Wall* (Dalsgaard & Halskov, 2010)—or remote controlled by other elements in the public space functioning as consoles—as in the case of *LOL* (Hespanhol, 2016).

9.6.2 The Impact of Context and Content

Another aspect that emerges from the preceding discussion is the relationship between the urban context—augmented and redefined by the addition of digital content—and the expectations for the interactivity of the resulting urban interface. In that regard, it is important to highlight the dual role of digital content in public spaces, which can both (1) add new functions and objective features to an urban setting and (2) offer the opportunity for new interactive and social experiences between strangers regularly sharing the same urban precinct—or *familiar strangers*, as Paulos and Goodman (2004) referred to them. As examples of adding new functions and features to an urban setting, we have the digital polling interfaces for citizen consultation in public spaces studied by Hespanhol and Tomitsch (2018), *Vote With Your Feet* (Steinberger et al., 2014), *Smart Citizen Sentiment Dashboard* (Behrens et al., 2014), and other related research on urban prototyping and digital pop-ups for community engagement (Fredericks, Tomitsch, Hespanhol, & McArthur, 2015; Hespanhol et al., 2015; Schroeter et al., 2012; Vlachokyriakos et al., 2014). In those studies, when encountering novel digital interfaces in public spaces, passers-by tended to fall back into ways of interacting inspired by familiar situations experienced elsewhere with interfaces involving similar functionality, input mechanisms, and feedback strategy, as demonstrated by the tests conducted by Hespanhol and Tomitsch (2018) when comparing surveys running on tablets and with an interactive audio device. Polling interfaces, by their nature, generally imply transferring explicit control of the input mechanism to a single user at a time, so that the vote can be counted or opinion expressed, therefore mostly falling back into the *operator* archetype, which is mostly unambiguous and intuitive.

Examples of new opportunities for new social interactions in public spaces are responsive installations running at public festivals, usually designed for quick interaction and often by many individuals simultaneously. To achieve this, they tend to rely on far fewer constraints regarding the input and feedback mechanisms and therefore depend on other factors such as the type and layout of the interactive public space to constitute cues toward their operation. For example, in the context of a street festival such as Vivid Sydney, where many of the light installations follow a straightforward interaction based on archetypes such as *shadow play*, *operator*, or *pet owner*, and in the absence of any clear interface affordances, visitors assume that some form of interaction may potentially be present in any digital media–based public setting and thus become open to exploration. In *Silent Island*, the mere possibility of creating shadows represented a way for visitors to appropriate some of the most obvious affordances of the installation for their own agenda—in this case, to have a playful experience in a public space with their family and friends.

9.7 CONCLUSION

In this chapter, I have described and discussed an expanded model for intuitive urban interfaces (Urbia). I have proposed a combination of some of the design variables initially proposed by Hespanhol and Tomitsch (2015)—notably regarding types of interactive public space and feedback strategies—with extra aspects related to public space layouts and input mechanisms, resulting in a set of contextual variables. Public space layouts refer to the nature of the sites as plazas (destinations) or thoroughfares (links between destinations). Input mechanisms correspond to the most common kinds of data entry and collection in public spaces—namely audio, consoles, mobile devices, or full-body interaction. The combination of different values assigned to those contextual variables thus results in different interactive city contexts—or urban interfaces—which in turn can be traced back to a series of corresponding interaction archetypes rooted in technology familiarity (Blackler, Mahar, et al., 2010) and the social dynamics unfolding in similar public spaces.

The resulting urban interfaces thus configure public space affordances supporting people's intuitive interpretation of the digitally enhanced built environment. On the downside, the inadvertent deployment of any of those combinations of contextual interfaces may lead, at best, to opportunistic appropriation of the interaction by the general public or, at worst, sheer confusion. To promote intuitive interaction with these types of interfaces, the design of plug-in architecture (Hespanhol & Tomitsch, 2017) constitutes a suitable strategy for curating the affordances of a public space and allowing thoughtful interventions by the general public into the responsive environment, which is then able to extend its features in accordance with familiar interaction archetypes—and, therefore, intuitively.

Admittedly, the Urbia Model here presented should not be taken as definitive, particularly in regards to the urban interaction archetypes proposed, which, although arguably reasonable, are surely far from exhaustive. Rather, it should be taken as a new iteration in better understanding the wide range of factors influencing—and often determining—the human interactions in our de facto new urban habitat.

REFERENCES

Alexander, C. A. (1977). *Pattern Language: Towns, Buildings, Construction*. Oxford University Press, New York, NY.

Antle, A. N., Corness, G., & Droumeva, M. (2009). Human–computer intuition? Exploring the cognitive basis for intuition in embodied interaction. *International Journal of Arts and Technology*, 2(3), 235–254.

Appadurai, A. (1991). Global ethnoscapes: Notes and queries for a transnational anthropology. In R. J. Fox (Ed.), *Recapturing Anthropology: Working in the Present*. Santa Fé, NM: School of American Research Press.

Behrens, M., Valkanova, N., Fatah gen Schieck, A., & Brumby, D. P. (2014). Smart citizen sentiment dashboard: A case study into media architectural interfaces. Paper presented at PerDis'14, Copenhagen, Denmark.

Blackler, A. (2008). *Intuitive Interaction with Complex Artefacts: Empirically-Based Research*. Saarbrücken, Germany: VDM.

Blackler, A., Desai, S., McEwan, M., Popovic, V., & Diefenbach, S. (2018). Perspectives on the nature of intuitive interaction. In A. Blackler (Ed.), *Intuitive Interaction: Research and Application* (pp. 19–40). Boca Raton, FL: CRC Press.

Blackler, A., & Hurtienne, J. (2007). Towards a unified view of intuitive interaction: Definitions, models and tools across the world. *MMI-Interaktiv, 13*(Aug.), 37–55.

Blackler, A., Mahar, D., & Popovic, V. (2010). Older adults, interface experience and cognitive decline. Paper presented at OZCHI 2010, "Design—Interaction—Participation," 22nd Annual Conference on the Australian Computer–Human Interaction Special Interest Group, Brisbane, Australia.

Blackler, A., Popovic, V., & Mahar, D. (2010). Investigating users' intuitive interaction with complex artefacts. *Applied Ergonomics, 41*(1), 72–92.

Blackler, A., Popovic, V., & Mahar, D. P. (2006). Towards a design methodology for applying intuitive interaction. Paper presented at the Design Research Society International Conference 2006, "WonderGround," Lisbon, Portugal.

Boring, S., Gehring, S., Wiethoff, A., Blöckner, M., Schöning, J., & Butz, A. (2011). Multi-user interaction on media façades through live video on mobile devices. Paper presented at CHI'11, Vancouver, Canada.

Brynskov, M., Dalsgaard, P., Ebsen, T., Fritsch, J., Halskov, K., & Nielsen, R. (2009). Staging urban interactions with media façades. Paper presented at INTERACT'09, Uppsala, Sweden, August 24-28.

Chattopadhyay, D., & Bolchini, D. (2015). Motor-intuitive interactions based on image schemas: Aligning touchless interaction primitives with human sensorimotor abilities. *Interacting with Computers, 27*, 327–343.

Dalsgaard, P., & Halskov, K. (2010). Designing urban media façades: Cases and challenges. Paper presented at CHI 2010, Atlanta, GA.

de Waal, M. (2014). *The City as Interface: How New Media Are Changing the City*. Rotterdam, the Netherlands: nai010.

Fischer, P. T., & Hornecker, E. (2012). Urban HCI: Spatial aspects in the design of shared encounters for media façades. Paper presented at CHI'12, Austin, TX.

Fredericks, J., Tomitsch, M., Hespanhol, L., & McArthur, I. (2015). Digital pop-up: Investigating bespoke community engagement in public spaces. Paper presented at the Annual Meeting of the Australian Special Interest Group for Computer–Human Interaction, December 07–10, Parkville, Australia: ACM.

Gamma, E., Helm, R., Johnson, R., & Vlissides, J. (1995). *Design Patterns: Elements of Reusable Object-Oriented Software*. Addison-Wesley, Reading, MA.

Hannerz, U. (1996). *Transnational Connections: Culture, People, Places*. New York: Routledge.

Hespanhol, L. (2016). Interacting with laughter: A case study on audio-based interactivity of public projections. Paper presented at OZCHI 2016, 28th Australian Conference on Computer–Human Interaction, Launceston, Australia.

Hespanhol, L., & Dalsgaard, P. (2015). Social interaction design patterns For urban media architecture. Paper presented at INTERACT'15, Bamberg, Germany.

Hespanhol, L., & Tomitsch, M. (2015). Strategies for intuitive interaction in public urban spaces. *Interacting with Computers, 27*(3), 311–326. doi.org/10.1093/iwc/iwu051.

Hespanhol, L., & Tomitsch, M. (2018). Power to the people: Hacking the city with plug-in interfaces for community engagement. In M. de Lange and Martijn de Waal (Eds.), *The Hackable City: Digital Media and Collaborative City Making in the Network Society*. Singapore: Springer.

Hespanhol, L., Tomitsch, M., Bown, O., & Young, M. (2014). Using embodied audio-visual interaction to promote social encounters around large media façades. Paper presented at DIS'14, Vancouver, Canada.

Hespanhol, L., Tomitsch, M., McArthur, I., Fredericks, J., Schroeter, R., & Foth, M. (2015). Vote as you go: Blending interfaces for community engagement into the urban space. Paper presented at C&T'15, Limerick, Ireland.

Hillier, B., & Hanson, J. (1984). *The Social Logic of Space*. Cambridge University Press, Cambridge, UK.

Hurtienne, J., & Blessing, L. (2007). Design for intuitive use: Testing image schema theory for user interface design. Paper presented at the 16th International Conference on Engineering Design, Paris, France.

Hurtienne, J., & Israel, J. H. (2007). Image schemas and their metaphorical extensions: Intuitive patterns for tangible interaction. Paper presented at TEI'07, New York.

Kokot, W. (2007). Culture and space: Anthropological approaches. *Ethnoscripts*, 9(1), 10–23.

Kray, C., Galani, A., & Rohs, M. (2008). Facilitating opportunistic interaction with ambient displays. Paper presented at CHI'08, Florence, Italy.

Lakoff, G., & Johnson, M. (1980). *Metaphors We Live By*. Chicago, IL: University of Chicago Press.

Lozano-Hemmer, R. (2001). Body movies. Relational Architecture 6. Retrieved January 13, 2015, from http://www.lozano-hemmer.com/body_movies.php.

Mueller, J., Walter, R., Bailly, G., Nischt, M., & Alt, F. (2012). Looking glass: A field study on noticing interactivity of a shop window. Paper presented at CHI'12, Austin, TX.

Norman, D. (2004). Affordances and design. Retrieved August 25, 2014, from http://www.jnd.org/dn.mss/affordances_and_desi.html.

Norman, D. (2013). *Design of Everyday Things: Revised and Expanded*. London: MIT Press.

Paulos, E., & Goodman, E. (2004). The familiar stranger: Anxiety, comfort, and play in public places. Paper presented at the SIGCHI Conference on Human Factors in Computing Systems, New York.

Schroeter, R., Foth, M., & Satchell, C. (2012). People, content, location: Sweet spotting urban screens for situated engagement. Paper presented at DIS'12, Newcastle, UK.

Steinberger, F., Foth, M., & Alt, F. (2014). Vote with your feet: Local community polling on urban screens. Paper presented at PerDis'14, Copenhagen, Denmark.

Tomitsch, M. (2016). City apps as urban interfaces. In A. Wiethoff & H. Hussmann (Eds.), *Media Architecture: Using Information and Media as Construction Material*. Berlin, Germany: De Gruyter.

Tomitsch, M., Ackad, C., Dawson, O., Hespanhol, L., & Kay, J. (2014). Who cares about the content? An analysis of playful behaviour at a public display. Paper presented at the International Symposium on Pervasive Displays, Copenhagen, Denmark.

United Nations. (2014). World's population increasingly urban with more than half living in urban areas. UN website, July 10. Retrieved June 7, 2017, from http://www.un.org/en/development/desa/news/population/world-urbanization-prospects-2014.html.

Vlachokyriakos, V., Comber, R., Ladha, K., Taylor, N., Dunphy, P., McCorry, P., & Olivier, P. (2014). PosterVote: Expanding the action repertoire for local political activism. Paper presented at DIS'14, Vancouver, Canada.

10 Designing Intuitive Products in an Agile World

Sandrine Fischer

CONTENTS

10.1 INTRODUCTION

Software development life cycles generally comprise various phases of design improvements, product development, and project management. One of the oldest development processes is called the *waterfall model*, under which professionals such as designers, engineers, and UX researchers operate in separate teams. Typically, development "flows" through successive stages: requirements specification, design, production, and maintenance. Success is measured by project completion at each stage—according to standards defined by each team—rather than the end results of the product. Because so much effort is spent on technical feasibility, cost, and timeframe of delivery, uncertainty can linger as to whether successive choices and implementations actually result in a product with a successful market life cycle.

This mindset is progressively being replaced by *agile* methodologies, whereby teams strive for user satisfaction and costumer success from day one. Rather than silo progress among disparate teams, the development of features, use cases, and attributes are listed as tasks in a *backlog* and prioritized per their business value or likelihood to meet end user needs. Product managers gather such tasks from stakeholders (e.g., product owners, engineers, user representatives) and then parse them into so-called sprints, or 2- to 4-week cycles of cross-functional teamwork (from specification to delivery, passing through design, development, and testing). By relying on user needs as a guiding mechanism, agile methodologies maximize the chance for products to compete within their landscape and achieve market growth. Modern organizations accomplish this by assembling highly cross-functional teams where members (e.g., engineers, developers, UX researchers, marketers) channel all operations and resources toward costumer end value.

Intuitive products have the potential to make a significant impact on today's app-saturated market; by improving first-time use, they lower the barrier for discovering what a product offers and eventually adopting it. But in what capacity can design and research geared toward intuitive use benefit agile teams? In this chapter, we discuss intuitive use from an operations perspective and examine the value it can bring to both teams and end users. Sections 10.2 and 10.3 review some caveats pertaining to implementing agile development and designing intuitive products, respectively. One takeaway worth noting is how the inner workings of intuitive behavior not only feed principles toward intuitive use design but also into innovation design. Section 10.4 maps these caveats onto a canvas we introduce for agile teams to factor intuitive use into their roadmaps. Section 10.5 presents a sprint-friendly implementation of the proposed canvas, and Section 10.6 discusses the value of considering intuitive use for team operations.

10.2 VALUES AND CHALLENGES IN AGILE SOFTWARE DEVELOPMENT

Along with agile methodologies, the current enterprise world is increasingly adopting practices known as *lean* software development. The first principle of agile development is customer satisfaction through the early and continuous delivery of valuable software (Beck et al., 2001). On the other hand, the first principle of lean development considers every activity or process that does not add some value to the company or its customers to be wasteful (Poppendieck and Poppendieck, 2003). Together, they advocate that any operation that does not clearly contribute to customer value should be considered wasteful (Alahyari, Svensson, & Gorschek, 2017). Thus, teams that strive to be both agile and lean must prioritize features that maximize end-user value and cut out any extraneous endeavors.

One assumes organizations that are agile would place user testing at the core of their operations. After all, what better way to surface the use cases and features that users value the most? In reality, practitioners have issues incorporating research that is perceived to be costly in terms of resources, verbose in terms of documentation, and not very insightful when it comes to informing project development. Through a series of surveys, interviews, and focus groups, Ardito et al. (2014) discovered that practitioners often skip usability evaluations on the reasoning that they are too resource demanding and difficult to implement due to a lack of suitable tools. Another issue was raised by Kashfi, Nilsson, and Feldt (2016)—namely, that is very difficult to align development operations teams with user-centric requirements. Taken together, these issues suggest that lean and agile teams expect methods that:

- Improve user experience, satisfaction, and product adoption
- Steer precise and impactful requirements
- Are sufficiently lightweight for 2- to 4-week sprints
- Can be validated and communicated convincingly to internal stakeholders

The rest of this chapter presents cognitive approaches to intuitive use that have the capacity to meet such criteria, starting by addressing some underpinnings and caveats.

10.3 CHALLENGES PERTAINING TO MAKING PRODUCTS INTUITIVE

10.3.1 DEVELOPMENT OF INTUITIVE BEHAVIORS

Intuition is a cognitive mode governing situations that, albeit previously unencountered, are resolved without conscious effort or clear after-the-fact justification (Epstein, Pacini, Denes-Raj, & Heir, 1996; Kahneman, 2003). Paradigms from cognitive psychology hold a long record of explaining how such intuitive behaviors develop. In artificial grammar learning, participants are exposed to sequences of characters that follow made-up rules (e.g., $U \rightarrow W \rightarrow C \rightarrow M$). Despite not being told about the rules, nor being prompted to learn the sequences, participants exhibit learning that is not only generative but also implicit (Reber, 1969). They are able to classify whether new sequences belong to the grammar or not, beyond the level of chance, but their explicit knowledge is insufficient for explaining their performance.

A widely held view is that the learning of a grammar—and likely relational structures among stimuli in general—stems from the detection and storage of the most frequently co-occurring bigrams and trigrams (for a review, see Pothos, 2007). Such a mechanism, which resembles the method of Bayesian inference, is more economical than storing (or hardcoding) each string separately. As more grammatical sequences are encountered, knowledge of short co-occurrences graduates to higher levels of complexity. The mechanism at stake, called *chunking*, leads to the assimilation of longer string fragments and configurations. Similarly, much of the structure behind complex configurations of stimuli that make up our environment is assimilated through the incidental and recursive chunking of frequent co-occurrences.

Through repeated mobilization, structural representations are increasingly reinforced to the detriment of the stimuli they embed, forming a type of knowledge representation called a *schema* (Bartlett, 1932; Rumelhart & Ortony, 1977). In schema representations, the structure common across instances is reinforced and fixed, while attributes that differ become optional/variable. This lifts some constraints regarding what can activate schemata, thereby facilitating their transfer to novel situations and contexts based on their structure (Gentner, Loewnestein, & Thompson, 2003; Gick & Holyoak, 1983; Reeves & Weiberg, 1994). This also prevails in artificial grammar learning, where participants can successfully map previously learned grammars not only onto new character sets but also onto completely different stimuli such as aural tones and colors (Altmann, Denies, & Goodes, 1995; Gomez, 1997; Kürten, DeVries, Kowal, & Zwitserlood, 2012). The structure mapping that is characteristic to schemata skips costly "bottom-up" and integrative processing. While low-level schemata override the processing of elementary stimuli, they eventually become embedded into higher-order schemata via chunking and are in turn overridden by these. To take a figurative example by Rumelhart and Ortony (1977), we see faces as configurations of eyes, a nose, a mouth, and so on, rather than an enormously complex configuration of shapes, colors, and shading. Higher-order schemata can make or break bottom-up thinking. As an example of the latter, Pothos (2005) tested a grammar of the routes traveled by a sales representative between cities in England (e.g., Birmingham → Leeds → Norwich → Sheffield). Learning appeared to be compromised when the grammar countered the intuitive expectation that sales representatives would plan routes efficiently by linking nearby cities first.

In sum, prior schemata are pervasive in our trains of thought, manifesting themselves as nano-moments where prior knowledge is applied without being prompted or formally instructed. Optional/variable values allow schemata to be transferred to situations and contexts previously not encountered. As more instances are compiled, idiosyncrasies are smoothed over (see schemata operations such as abstraction, induction, and progressive alignment; Gick and Holyoak, 1983; Jung and Hummel, 2015; Kandaswamy et al., 2014; Reber, 1969), making it difficult to impute schemata with a single instance in particular. Finally, the structure mapping they operate overrides the analysis of lower-level attributes, rendering thinking automatic and effortless.

10.3.2   CAUTIONARY PRINCIPLES OF INTUITIVE USE DESIGN

When prior schemata substitute themselves into situational analyses, the automatic thinking they elicit is at the expense of encoding content and details. Information that is highly schema-compatible thus results in fewer situational cues being encoded, reaching awareness, or leaving a trace in episodic memory. In a now classic demonstration by Brewer and Treyens (1981), university students were made to wait in a university office and later asked to recall the objects present in the room. Participants barely recalled objects that were not consistent with the schema of an office, such as a brick. In addition, they incorrectly recalled objects highly compatible with the schema of an office that were *not* present (e.g., books). Such distortions scale to complex behaviors, raising two Cautionary Principles that design teams should consider in the early stages of the development cycle.

Consider what would happen if an interior designer were tasked to upgrade the university office. Since the addition of objects consistent with an office would not increase the chance of activating the schema, adding books (say) would not make the office more intuitive to the participants. Enter the First Cautionary Principle (CP1): *Because situations highly compatible with prior schemata entail transfer that overrides low-level detail, they do not necessitate interventions that are explanatory or instructional.* In other words: If it isn't broke, don't fix it. Conversely, the addition of objects neutral or non-consistent with an office would not prevent the schema from being activated. Indeed, the interior designer could add a brick to the room without reducing any intuition that the room is an office. Thus, a corollary Second Cautionary Principle (CP2) emerges: *Common situations need not be stereotypical; they are already resilient to restyling and can be made more innovative without sacrificing understanding.* Here, restyling is any effort to introduce creativity, originality, or innovation into a product's aesthetic or interaction design.

This notion of redesign not working with schema-compatible content stems from Cognitive Load Theory and Instructional Systems Design (see Fischer et al., 2015b). Numerous experiments in those fields show that crafting task environments with instructions, training, or simplifications only benefits participants who lack the relevant schemata. When participants already possess domain knowledge, though, the same interventions have no adverse impact (Kalyuga, 2007; Mayer, 2001). Currently, the dominant belief in the intuitive use community is that designing features with more familiarity makes a product more intuitive (Blackler & Hurtienne, 2007; Hurtienne, Horn, Langdon, & Clarkson, 2013). Fischer, Itoh, and Inagaki (2015b), who cautioned this is not the case for applications that already tap into existing knowledge (i.e., are very common), discuss several examples of intuitive use redesign that show no improvement due to the devices in question being very common. In one case, the redesign of a ticket-vending machine through metaphors, familiar tabs, clear labels, and so on improved neither overall usage nor satisfaction (Hurtienne et al., 2013). In another, a simplified menu structure did not improve error rates during first-time use of a pet-sitting game, although it did improve intuitive uses as coded by the researchers (Gudur, Blackler, Popovic, & Mahar, 2013), while the addition of various familiarity levels to the features of a microwave oven did not improve correct uses (Blackler, Popovic, & Mahar, 2014). Because they did not improve first-time performance but did not degrade these performances either, the three aforementioned redesigns strike one as instantiations of CP2.

This is not always the case, though: interventions on schema-compatible content can create saliencies that disengage users from the intuitive processing granted by their schemata. The consequence is unwanted sequential or analytical thinking, with all the cognitive load and behavioral divergence this incurs. This issue, called the *expertise reversal effect* in Instructional Systems Design, may explain why adding icons to text labels worsened first usage among seniors instead of improving intuitiveness (Gudur, Blackler, Popovic, & Mahar, 2014). Framing intuitive use in terms of schema transfer exposes critical trade-offs for what can be achieved through design. We define the effectiveness of (re)design (Δ in Figure 10.1) as the difference in performance before and after an intervention. Figure 10.1, which extends a previous illustration from Fischer et al. (2015b), portrays Δ as a function of where features lie in the users' schema space.

In this view, products and features that are incompatible with prior schemata are more likely to benefit from interventions tailored toward instruction (CP1). However, features already compatible with prior schemata, such as those for common household appliances, are unlikely to benefit from instructional design intervention; in fact, such interventions may even be counterproductive. However, such features should be resilient to restyling, creativity, or differentiation (CP2). In the following section, we introduce a canvas for applying both Cautionary Principles onto the planning and development of products.

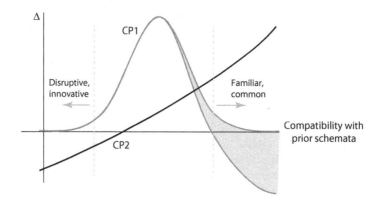

FIGURE 10.1 Illustration of the non-linear effect of design interventions. The vertical axis, labeled Δ, is the relative change in performance before and after the intervention, or how effective (re)design is. Low compatibility (left) includes disruptive use cases and innovative technology, while high compatibility (right) includes familiar use cases and common appliances. In the latter case, instructional interventions range from ineffective to counterproductive, indicated by a spread in Δ for CP1.

10.4   CANVAS FOR DEVELOPING INTUITIVE PRODUCTS IN AGILE SETUPS

To most teams, the principles outlined in Figure 10.1 are likely to be challenging to turn into a product roadmap. An emerging practice for organizations facing complex development is to turn toward management media such as *canvases*. The prototypical canvas was designed by Osterwalder (2008) in order to model and deliver business value. Since then, other canvases have appeared for niche areas such as user-centered design or lean development. These template tools enable cross-functional teams to deliver end user value by mapping development activity onto key performance indicators.

The Intuitive Product Development (IPD) canvas introduced here is the first effort to align the development of new features in terms of users' prior schemata. For those new to agile development, *sprint zero* is the preliminary stage whereby stakeholders shape the product's vision and outline the backlog for delivery (Cajander, Larusdottir, & Gulliksen, 2013; Larusdottir, Cajander, & Gulliksen, 2013; Larusdottir, Cajander, & Gulliksen, 2012). The IPD canvas is particularly well suited for the sprint zero phase, where lo-fi prototype testing can inform the development plan without reducing delivery velocity. While the proposed canvas does not address business value directly, it tackles the difficulties associated with first-time usage and onboarding experience. The first three activity blocks of the IPD canvas, templated in Figure 10.2, are fairly common to user-centered design: *Requirements*, *Lo-Fi Prototyping*, and *Evaluation*. A fourth activity block focuses on backlog prioritization and sprint planning for upcoming *n*-week development cycles.

The Requirements block is used to inventory features and use cases related to the product. At a minimum, the inventory is created by meeting with stakeholders such as product owners, engineering leads, and user representatives. Improvements

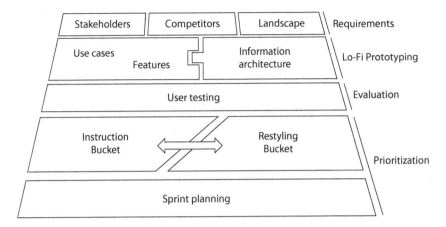

FIGURE 10.2 Intuitive Product Development (IPD) canvas. The double arrow indicates that the proportion of features belonging to each bucket depends on the overall level of innovation (vs. commonality) of the product.

can be made by surveying competitor products, technical requirements, and the extended business landscape. Later in this chapter, a demonstration ("demo") will expand on why the Requirements block calls for more thoroughness and resources than the subsequent Lo-Fi Prototyping block. The sole purpose of prototyping is to translate the feature space and product assumptions into testable requirements. As far as the canvas concerns itself with user schemata, this activity can be limited to a comprehensive structuring of the information architecture, along with the labeling of product features in a menu–submenu fashion. The information architecture should be tested in the Evaluation block with methods that assess the schema compatibility of each feature (see Section 10.4.1 onward for details). During the ensuing *Prioritization* block, features are dispatched into one of two *buckets* corresponding to either Cautionary Principle:

- *Instruction Bucket (CP1):* Features incompatible with schema should be placed in this bucket because they are prone to human error. Design teams should prioritize iterations for these features in order to reduce their deleterious impact on first-time performance, usability, and the onboarding experience. This bucket calls for techniques tailored toward explanation, tutorials, and the simplification of features/attributes, possibly through improved labels, "infotips," interactive tutorials, or even the removal of the feature altogether.
- *Restyling Bucket (CP2):* Features compatible with the schema should be placed in this bucket and iterated later in the development cycle. Teams may take advantage of the resilience of these features by making them creative, adding innovation, or differentiating them from the competition.

The gains (Δ in Figure 10.1) and metrics to monitor when iterating on (i.e., redesigning from) each bucket follow the same divide. One can expect iterations on the Instruction

Bucket to boost perceived usability (e.g., traits such as ease of understanding, learning, or use) versus enjoyment or delight for the Restyling Bucket. But because subjective appraisal often fails to catch objective changes, behavioral metrics should come first. Products that tap into novel schema spaces tend to foster human errors as they force first-time users into task solving and deliberation, so performance is best reflected by task accuracy (Camp et al., 2001). Conversely, products that tap into a common knowledge space benefit from the transfer of existing schemata, meaning their first-time use leans toward the preconscious/intuitive. First-use performance of such products is better measured through metrics such as decision times, completion times, and automaticity (e.g., dual-task performance; see discussion in Fischer et al., 2015b). Because products made of a mixture of common and novel features involve task solving, they should primarily be assessed by accuracy metrics; for example, redesigning the product should result in fewer usage errors and task failures.

10.4.1 Quantitative Methods toward Diagnosing User Schemata

The IPD canvas revolves around an understanding of where each feature and use case lies in the schema space of an audience. However, there is a catch: knowledge is a multilevel assembly of schemata, and transfer depends on variable optionality and representation abstractedness. To wit, it often occurs that familiar situations fail to trigger transfer, and seemingly novel situations elicit transfer by mapping to a schema inherited from a remote task domain. These challenges associated with the schema construct make a strong case in favor of cognitive experimentation.

In this vein, Fischer, Itoh, and Inagaki (2009) posited that schema compatibility elicits a *transfer operation* that can be experimentally separated from converse operations such as *induction*, whereby new schemata are formed. The occurrence of these two types of operations in first-use performance can be operationalized with a *schema induction paradigm*. In a nutshell, participants are placed into groups and asked to undergo a manipulation of the schemata regarding the interface they are about to use. The procedure, detailed in Fischer, Itoh, and Inagaki (2015a), entails the following:

- A *control group*, who use the interface with nothing but their prior schemata
- A *reading group*, who are asked to superficially study the interface, thereby amending their prior schemata
- An *induction group*, who perform tasks designed to induce new schemata related to the interface being tested

The three groups then undergo the same usage scenario, and their performances are logged and compared at the granularity of the interface's features through effect size calculations (Fischer, Itoh, & Inagaki, 2014). Following this quantification, several behavioral patterns can be isolated.

- *Inoperative induction*, corresponding to features that participants failed to learn (equally large errors for all groups) and that should be placed in the Instruction Bucket

- *Positive induction*, corresponding to features that participants learned the most (largest effect toward schema induction) and that should also be placed in the Instruction Bucket
- *Transfer*, corresponding to features that are intuitive (very few errors for all groups) and that should be placed in the Restyling Bucket

While Fischer et al.'s (2015b) *screening method* operationalizes the influence of schemata on behavior (viz. first-usage performance), other paradigms exist in cognitive psychology that could be used to operationalize other aspects of the schema construct. The propensity of schemata to automate processing leads to dual-task experiments, especially in cases where a rather automatic activity (e.g., driving a car) requires additional cognitive load (driving while using an onboard computer or having a phone conversation, etc.; Engstrom, 2011; Rasmussen, 1987; Summala, 1988; Zhou, Itoh, & Inagaki, 2008). Another interesting property is the intrusions schemata cause to episodic memories—namely, memories of past events that include contextual information about a place, moment, detail, or internal thoughts present at that time. Brewer and Treyens (1981) revealed that schema-compatible information is overlooked; hence, computational resources can be freed and rerouted toward information atypical for the background schema. In the short term, atypical information is better remembered than information that is consistent with prior schemata and thus neutral. However, the episodic trace associated with atypical information decays over time (Bartlett, 1932), whereas background schemata are here to stay.

False memories are operationalized in lab settings by the Deese–Roediger–McDermott (DRM) paradigm (Roediger & Mc Dermott, 1995), recently adapted by Lee and Ryu (2014) for the purpose of testing products. In a typical DRM experiment, participants study a list of words (e.g., hot, winter, etc.) related to a set of undisclosed words (e.g., cold). The latter serve as *lures* in a follow-up test where participants must recognize (or recall) the presented words. In Lee and Ryu's *False Belief Technique*, participants were asked to

- Use a device, in this case a multifunctional smart TV,
- Take a test one week later asking whether they recognized features encountered in the device among consistent lures, indicating their confidence on a six-point Likert scale.

The two experimental conditions (listed vs. lure features) allowed the correctness, full confidence, and mean confidence of recognition to be calculated and summed into so-called appropriateness values for both listed and lure features. Following this quantification, features with the greatest appropriateness scores are most consistent with participants' schemata. Following CP2, such features should be good candidates for the Restyling Bucket.

Research is underway to fully scope the construct validity of the screening method and the False Belief Technique. What is of more concern to this chapter is ecological validity and agility. The former addresses the capacity of such methods to enact (re)designs under the Cautionary Principles of the IPD canvas, while the latter relates to their sprint-friendliness.

10.5 SPRINT DEMO

Cognitive paradigms place a great emphasis on scripting to limit the exchanges between participants and the experimenter. The fact that such methods require little to no moderation makes them great candidates for remote (or online) testing. Contrary to in-lab testing, remote testing is asynchronous and unmoderated, meaning participants are not guided by a test moderator, which saves a great deal of logistical effort. The research behind this sprint demo is a push in this direction.

All pretests and tests took place online using the following software: Amazon Mechanical Turk to recruit participants, Google Forms to administer instructions and questionnaires, Testable (2017) to run the schema methods, Treejack (2017) to implement and test the information architecture, and modules of URU (a platform for unmoderated and remote user testing; Fischer, Itoh, & Inagaki, 2015c) to screen the schemata operations. This section presents the test flow for a diagnosis that combines the screening method and False Belief Technique, the insights garnered, revisions that were made, and the outcome. The full procedure entails pretests for technicalities beyond the scope of a sprint demo. The quality of crowd-sourced data is typically inferior to data collected in a lab under supervision. As Stoycheff (2016) suggests, this can be mitigated by amplifying the social exchange dimension of the procedure. In our case, this entailed a scenario more meaningful to participants, instructions that were short and crystal clear, and interim checks of response quality.

10.5.1 REQUIREMENTS, PROTOTYPING, AND TESTING

The prototype is derived from DoIT#, an onboard vehicle computer previously developed following the requirements and prototyping phases of Figure 10.2. The information architecture and functions of DoIT# had been sourced from competitor and landscape surveys, and its labels and menu logic had been refined with user interviews. Onboard computers are comprised of various technologies that are not equal in terms of innovativeness. In line with the IPD canvas's global stance, priority should be given to testing the most advanced technological offerings in such products. In the case of DoIT#, this meant focusing on features for GPS navigation and onboard computer menus, as opposed to those for air conditioning or music. Those corresponded to a subset of 11 states (i.e., distinct screens of an interface), with between three and seven features (either menu entries or functions) in each.

In order to diagnose the resulting prototype in terms of schemata operations and feature appropriateness, each state was pruned of half of its features:

- One half remained in the information architecture that underwent the screening method.
- The other half was reserved to serve as lures for the False Belief Technique.

Conforming to the *Lo-Fi Prototyping* advocated in Section 10.4, the sprint worked with a bare information architecture of the prototype, implemented in Treejack (Figure 10.3a), instead of the GUI-rich version of DoIT# tested in earlier research (Fischer et al., 2009, 2015a, 2015b).

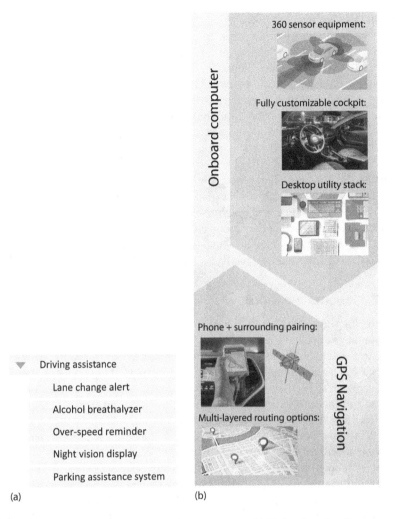

(a)

(b)

FIGURE 10.3 (a) Screenshot of a state of the prototype. (b) Technology brochure.

Sixty-nine Mechanical Turkers took part in a 15-minute study designed after the screening method and were remunerated 4 USD. Since onboard computers are used by adults of all demographics and backgrounds, no specific recruitment criteria were enforced other than participants being over 18 years old and residing in the United States. Participants were assigned either to a *control* ($N=20$), *reading* ($N=24$), or *induction* group ($N=25$). The reading and induction groups studied screenshots of the prototype (Figure 10.3a) under schema-amending and -induction conditions, respectively (Fischer et al., 2015a); the control group skipped this step. The three groups then interacted with the prototype under the following scenario:

> You just requested the route to your friend's house. First, set up this route so that it avoids tolls (task 1). Then, display the navigation directions in the rearview mirror (task 2). Finally, activate the assistance system for passing other cars (task 3).

Each task was displayed sequentially, and participants navigated the information architecture until they found a feature they deemed fulfilled it. They could then proceed at their own pace to the next task. Effect sizes were calculated by averaging the number of clicks per feature. Following the analysis, features displaying induction patterns were placed in the Instructional Bucket, and features displaying transfer patterns were placed in the Restyling Bucket.

One week later, participants from the reading and induction groups were invited back for a four-minute study designed after the False Belief Technique and remunerated 3 USD. The responding participants (10 from reading, 14 from induction) were given 31 recognition trials, 20 of which involved features from the prototype, the rest being the lures initially pruned from DoIT#. Prototype features (not lures) had been shown in the screenshots that both groups had studied in amendment versus induction conditions. Features and lures were randomly assigned to trials, and participants indicated whether they had seen them or not in the prototype and rated the confidence of that answer. Features and lures were sorted from the most to the least appropriate, according to Lee and Ryu's formula (2014).

10.5.2 INSIGHTS AND REDESIGN

The screening method identified a number of features to be addressed in the next sprint. Some features ("Driving indices," "Guidance clues," and "Map") displayed an inoperative induction pattern, in that participants failed to induce the corresponding schemata. Such features were relabeled for clarity (e.g., "Map" became "Map view"; "Driving indices" became "Cruising data"). Other features ("Onboard computer," "Driving assistance," "Lane change alert," "Information and alerts," "Miscellaneous," "Displays," and "Rearview mirror") displayed a strong induction pattern: namely, participants initially lacked and effectively induced the corresponding schemata. These gaps were addressed by creating a five-vignette brochure designed to disambiguate the DoIT# onboard computer and navigation technology offering (Figure 10.3b).

The "360 sensor equipment" vignette was intended to clarify that onboard computers assist driving through sensor grids that are independent from the GPS. This fix had already been identified in a previous screening of DoIT# with respect to knowledge schemata of Japanese university students (Fischer et al., 2015b). The "Desktop utility stack" was intended to clarify the "Miscellaneous" menu by implanting the notion that onboard computers can propose basic desktop utilities (agenda, currency converters, etc.). As it was designed to support high-level understanding without mentioning specific clues about the information architecture itself, the brochure did not reuse the same feature labels.

The False Belief Technique revealed some lures of high appropriateness to the participants, such as "Gas autonomy," "Average speed," "Bar/pascal," and "Adjust to traffic." Expanding on one of these lures, participants judged the phrase "Bar/pascal" without any context about tire pressure and yet were highly confident it was part of the prototype. This indicates the high compatibility of this feature with participants' prior schemata, resulting in the belief that the prototype expressed tire pressure in units of bars/pascals. Following this insight, the highly appropriate lures were reintegrated into the prototype at their original location within the DoIT# information architecture. In summary, the revised prototype had three features relabeled and five added, and was supplemented by the technology brochure (Figure 10.3b) regarding what to expect from such a product.

10.5.3 REDESIGN OUTCOMES

A follow-up study was conducted to assess the outcome of the aforementioned revisions. A new pool of Turkers underwent the same usage scenario as in the diagnosis phase, either with the original ($N = 17$) or revised prototype (preceded by the brochure; $N = 20$). Participants were given several questionnaires to gauge the potential subjective impact of the revisions. They were asked to rate their experience with the prototype using seven-point Likert scales. Other than "Intuitive," the descriptors ("Confusing," "Practical," "Premium," "Complicated," "Innovative," "Unpredictable") were selected from AttrakDiff, an instrument for conducting longitudinal evaluations of user experience (Hassenzhal, Burmester, & Koller, 2008). They were also asked, when renting a car for business, if they would want an onboard computer with more, less, or similar technology. Finally, they rated how likely they would "recommend such an onboard computer to a friend or colleague" on a 10-point Likert scale (1 for "Not at all," 10 for "Absolutely"). This metric, commonly known as the Net Promoter Score (NPS; Reichheld, 2003), is popular in industry for tracking product loyalty over time.

Figure 10.4 summarizes the main outcomes of this study. First-use performance registered a two-fold improvement, from 23.5% of participants correctly completing the full scenario with the original prototype to 50% with the revised prototype. The revised prototype was technologically less intimidating, with 20% of the participants stating they would rent a car with less technology, compared with 33% for the original. The NPS was calculated by subtracting the percentage of *detractors*, or participants who responded between 1 and 7, from the percentage of *promoters*, or those who responded 9 or 10. Interestingly, this cutoff is reminiscent of the 3:1 ratio for negative-to-positive emotions proposed by Fredrickson (2004). This controversial metric gained success in the industry with teams such as Android at Google (Garb and Roeber, 2013), who used it to communicate their user-centered design efforts. The NPS score ranges from −100 (all respondents are detractors) to +100 (all are promoters). An NPS above 0 is deemed good; one above +50 is deemed excellent. For reference, the median NPS originally surveyed for 400 large companies was +16 (Reichheld, 2003). The NPS for this redesign increased from −0.3 for the original prototype to +12 for the revised one. The revisions did not move the needle for average ratings of AttrakDiff-like descriptors, albeit the revised prototype marginally stood out as more premium and innovative than the original one. One interesting path of research would be to determine whether the discrepancies between objective performance and subjective experience are due to participants devoting much of their cognitive resources to means – end analysis, leaving few resources available for awareness of their experience. In this sense, and in line with the Δ criteria discussed in Section 10.3.2, experience ratings for redesigns of common products would be more differentiated than those for innovative ones, and vice versa for first-usage performance.

As previously acknowledged, such a demo is not intended to fully validate the methods. Nevertheless, it illustrates how CP1 and CP2 can be leveraged to inform the

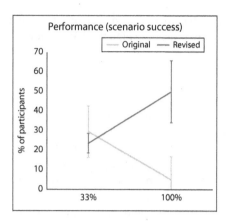
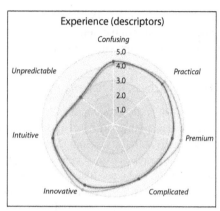

FIGURE 10.4 Redesign outcomes.

product development cycle. The revisions pertaining to CP1 were fairly basic. Still, it seems that applying them to features identified via the screening method improved first-usage performance. The takeaway here is that relabeling the majority of features would be time-consuming for designers, yet doing so for the subset of features that needs it the most adds value both in terms of team operations and end-user value. Revisions pertaining to CP2 involved adding schema-compatible features compliant with the False Belief Technique. Because an onboard computer prototype is technologically challenging, CP2-related work should commence only after an acceptable level of intuitiveness has been reached for features in the Instruction Bucket. Additional functionality is not the only way to address features in the Restyling Bucket. Features that make it to this bucket (viz. their schema compatibility is established) can be reworked along the hierarchy of design layers portrayed by Garrett (2011) or Norman's Action Cycle (1988). Such features could be restyled in spite of an original interaction logic, command, and/or visual design.

10.6 CONCLUSION

Most practitioners may not think of intuitive use as a necessary requirement. And yet, this chapter posits that the topic of intuitive use, and its implementation, is largely relevant and worth investing in. Intuitive use is not solely about improving first usage of products or fostering a seamless onboarding experience. Recent developments that link findings in Intuitive Interaction Design and Instructional System Design (Section 10.3.2; see also Fischer et al., 2015b) reveal that intuitive use is equally relevant for fueling differentiation and innovation in products. The core argument is that intuitive use is rooted in knowledge schemata, and this framework can help to avoid many conceptual and methodological pitfalls. One such pitfall concerns the non-linear effect of design interventions with respect to prior schemata. As such, enhancing task environments with familiar cues does not guarantee intuitiveness per se. One strategy is to diagnose prior schemata and optimize task environments accordingly—an endeavor in line with modern paradigms of agile and lean development.

Another pitfall deals with the elusiveness of knowledge schemata. Several cognitive paradigms could help to tackle this issue, potentially contributing to their acceptance in the practitioner community. The sprint demo in this chapter showcases two such methods. Fischer et al.'s screening method and Lee and Ryu's False Belief Technique are practical implementations of implicit (indirect) and explicit (direct) tests, respectively. The former captures the influence of schemata on behavior, while the latter accounts for their influence on memory self-reports. The IPD canvas enables either implementation to be articulated during product testing. While methods can be fine-tuned in the future, the sprint demo contributes a demonstration of their practicality and ecological validity in unmoderated settings.

The major takeaway is that design media as simple as naked information architectures are sufficient for screening candidate features in terms of schemata operations. The screening method informed revisions that were significant in terms of end user value: twice as many participants were able to complete the prototype onboarding, thus gaining a better sense of what the product offered. The False Belief Technique additionally offered insights as to how the information architecture could be further populated at very low cost. The IPD canvas, along with the additional diagnostic methods proposed in this chapter, enables agile teams to jumpstart their sprint zero and prioritize their operations before any development is actually carried out. The bottom line is that by applying schema methods to an information architecture, features spanning a large opportunity space (competitor, landscape) can be efficiently mapped out and prioritized for development.

REFERENCES

Alahyari, H., Svensson R. B., & Gorschek. T. (2017). A study of value in agile software development organizations. *The Journal of Systems & Software*, *125*(C), 271–288.

Altmann, G. T. M., Denies, Z., & Goodes, A. (1995). Modality independence of implicitly learned grammatical knowledge. *Journal of Experimental Psychology: Learning, Memory and Cognition*, *21*(4), 899–912.

Ardito, C., Buono, P., Caivano, D., Costabile, M. F., & Lanzilotti, R. (2014). Investigating and promoting UX practice in industry: An experimental study. *International Journal of Human–Computer Studies*, *72*(6), 542–551.

Bartlett, F. C. (1932). *Remembering: A Study in Experimental and Social Psychology*. Cambridge University Press, Cambridge, UK.

Beck, K., Grenning, J., Martin, R. C., Beedle, M., Highsmith, J., Mellor, S., van Bennekum, A., et al. (2001). Manifesto for agile software development. Agile Alliance. Online: http://agilemanifesto.org/principles.html (accessed September 2017).

Blackler, A., & Hurtienne, J. (2007). Towards a unified view of intuitive interaction: Definitions, models and tools across the world. *MMI-Interaktiv*, *13*, 36–54.

Blackler, A., Popovic, V., & Mahar, D. (2014). Applying and testing design for intuitive interaction. *International Journal of Design Sciences and Technology*, *20*(1), 7–26.

Brewer, W. F., & Treyens, J. C. (1981). Role of schemata in memories for places. *Cognitive Psychology*, *13*(2), 207–230.

Cajander, Å., Larusdottir, M. K., & Gulliksen, J. (2013). Existing but not Explicit–The user Perspective in Scrum Projects in Practice. In: Kotzé P., Marsden G., Lindgaard G., Wesson J., & Winckler M. (Eds). INTERACT 2013. Lecture Notes in Computer Science. Springer, Berlin, Heidelberg, vol. 8119, pp. 762–779.

Camp, G., Paas, F., Rikers, R., & van Merrienboer, J. (2001). Dynamic problem selection in air traffic control training: A comparison between performance, mental effort and mental efficiency. *Computers in Human Behavior, 17*, 575–595.

Engstrom, J. (2011). Understanding attention selection in driving: From limited capacity to adaptive behavior. PhD thesis, Chalmers University of Technology, Gothenburg, Sweden.

Epstein, S., Pacini, R., Denes-Raj, V., & Hier, H. (1996). Individual differences in Intuitive-Experiential and Analytical-Rational thinking styles. *Journal of Personality and Social Psychology, 71*(2), 390–405.

Fischer, S., Itoh, M., & Inagaki, T. (2009). A cognitive schema approach to diagnose intuitiveness: An application to onboard computers. In *Proceedings of the 1st International Conference on Automotive User Interfaces and Interactive Vehicular Applications*, Essen, Germany, pp. 35–42.

Fischer, S., Itoh, M., & Inagaki, T. (2014). Identifying the cognitive causes of human error through experimentation. *European Journal of Automation, 48*, 319–338.

Fischer, S., Itoh, M., & Inagaki, T. (2015a). Prior schemata transfer as an account for assessing the intuitive use of new technology. *Applied Ergonomics, 46*, 8–20.

Fischer, S., Itoh, M., & Inagaki, T. (2015b). Screening prototype features in terms of intuitive use: Design considerations and proof of concept. *Interacting with Computers, 27*(3), 256–270.

Fischer, S., Oelkers, B., Fierro, M., Itoh, M., & White, E. J. (2015c). URU: A platform for prototyping and testing compatibility of multifunction interfaces with user knowledge schemata. In *Proceedings of the Human–Computer Interaction International Conference 2015*, Los Angeles, CA, pp. 151–160.

Fredrickson, B. L. (2004). The broaden-and-build theory of positive emotions. *Philosophical Transactions of the Royal Society B: Biological Sciences, 359*(1449), 1367–1378.

Garb, R., & Roeber, H. (2013). Enchant, simplify, amaze: Android's design principles. *Google I/O 2013*, San Francisco, CA. Online: https://developers.google.com/events/io/2013/sessions (Accessed September 2017).

Garrett, J. J. (2011). *The Elements of User Experience: User-Centered Design for the Web and Beyond, 2nd Ed.* New Riders Publishing Thousand Oaks, CA.

Gentner, D., Loewenstein, J., & Thompson, L. (2003). Learning and transfer: A general role for analogical encoding. *Journal of Educational Psychology, 95*(2), 393–408.

Gick, M. L., & Holyoak, K. J. (1983). Schema induction and analogical transfer. *Cognitive Psychology, 15*, 1–38.

Gomez, R. L. (1997). Transfer and complexity in artificial grammar learning. *Cognitive Psychology, 33*(2), 154–207.

Gudur, R. R., Blackler, A. L., Popovic, V., & Mahar, D. P. (2014) Adaptable interface model for intuitively learnable interfaces: An approach to address diversity in older users' capabilities. In Y.-K. Lim, K. Neidderer, J. Redström, E. Stolterman, & A. Valtonen (Eds.), *DRS 2014: Design's Big Debates Pushing the Boundaries of Design Research*, Umea Institute of Desgin, Umea University, Umea, Sweden, pp. 374–387.

Gudur, R. R., Blackler, A., Popovic, V., & Mahar, D. (2013). Ageing, technology anxiety and intuitive use of complex interfaces. *Lecture Notes in Computer Science: Human–Computer Interaction; INTERACT 2013*, pp. 564–581.

Hassenzahl, M., Burmester, M., & Koller, F. (2008). Der User Experience (UX) auf der Spur: Zum Einsatz von www.attrakdiff.de. In: Brau, H., Diefenbach, S., Hassenzahl, M., Koller, F., Peissner, M., Röse, K. (Eds.) Usability Professionals, German Chapter der Usability Professionals Association, Stuttgart, Germany, pp. 78–82.

Hurtienne, J., Horn, A.-M., Langdon, P., & Clarkson, J. (2013). Facets of prior experience and the effectiveness of inclusive design. *Universal Access in the Information Society, 12*, 297–308.

Jung, W., & Hummel, J. E. (2011). Progressive alignment facilitates learning of deterministic but not probabilistic relational categories. In *Proceedings of CogSci2011*, Boston, MA, 33.

Jung, W., & Hummel, J. E. (2015). Making probabilistic relational categories learnable. *Cognitive Science*, *39*(6), 1259–1291.

Kahneman, D. (2003). A perspective on judgment and choice: Mapping bounded rationality. *The American Psychologist*, *58*(9), 697–720.

Kalyuga, S. (2007). Expertise reversal effect and its implications for learner-tailored instruction. *Educational Psychology Review*, *19*, 509–539.

Kandaswamy, S., Forbus, K., & Gentner, D. (2014). Modeling Learning via Progressive Alignment using Interim Generalizations. In Bello, P., Guarini, M. McShane, M. & Scassellati, B. (Eds.). Proceedings of the 36th Annual Meeting of the Cognitive Science Society, Austin, TX, pp. 2471–2476.

Kashfi, P., Nilsson, A., Feldt, R. (2017). Integrating User eXperience practices into software development processes: implications of the UX characteristics. PeerJ Computer Science, 3, e130.

Kürten, J., DeVries, M. H., Kowal, K., & Zwitserlood, P. (2012). Age affects chunk-based, but not rule-based learning in artificial grammar acquisition. *Neurobiological Aging* *33*(7), 1311–1317.

Larusdottir, M. K., Cajander, A., & Gulliksen, J. (2012). The big picture of UX is missing in Scrum projects. In *Proceedings of the Ix-Used Workshop at NordiCHI 2012*, Copenhagen, Denmark, pp. 49–54.

Larusdottir, M. K., Cajander, A., Gulliksen, J. (2013). Informal feedback rather than performance measurements: User centred evaluation in Scrum projects. *Behavioral Information Technology*, *33*(11), 1118–1135.

Lee, J., & Ryu, H. (2014). I remember/know/guess what i saw: A false belief technique to features selection. In *Proceedings of CHI'14: Extended Abstracts on Human Factors in Computing Systems*, Toronto, Canada, pp. 2593–2598.

Mayer, R. E. (2001). *Multimedia Learning.* New York: Cambridge University Press.

Osterwalder, A. (2008). The business model canvas. July 05. Online: nonlinearthinking.typepad.com (accessed September, 2017).

Poppendieck, M., & Poppendieck, T. (2003). *Lean Software Development: An Agile Toolkit.* Addison-Wesley Professional, Boston, MA.

Pothos, E. M. (2005). Expectations about stimulus structure in implicit learning. *Memory & Cognition*, *33*(1), 171–181.

Pothos, E. M. (2007). Theories of artificial grammar learning. *Psychological Bulletin*, *133*(2), 227–244.

Rasmussen, J. (1987). The definition of human error and a taxonomy for technical system design. In J. Rasmussen, K. Duncan, & J. Leplat (Eds.), *New Technology and Human Error* (pp. 23–30). Chichester, UK: Wiley.

Reber, A. S. (1969). Transfer of syntactic structure in synthetic languages. *Journal of Experimental Psychology*, *81*(1), 115–119.

Reeves, L. M., & Weiberg, R. W. (1994). The role of content and abstract information in analogical transfer. *Psychological Bulletin*, *115*(3), 381–400.

Reichheld, F. F. (2003). The one number you need to grow. *Harvard Business Review*, December. Online: http://hbr.org/2003/12/the-one-number-you-need-to-grow (accessed September 2017).

Roediger, H. L., & McDermott, K. B. (1995). Creating false memories: Remembering words not presented in lists. *Journal of experimental psychology: Learning, Memory, and Cognition*, *21*(4), 803.

Rumelhart, D. E., & Ortony, A. (1977). The representation of knowledge in memory. In R. C. Anderson, R. J. Spiro, & W. E. Montague (Eds.), *Schooling and the Acquisition of Knowledge* (pp. 99–135). Hillsdale, NJ: Erlbaum.

Stoycheff, E. (2016). Please participate in Part 2: Maximizing response rates in longitudinal MTurk designs. *Methodological Innovations*, 9, 1–5.

Summala, H. (1988). Risk control is not risk adjustment: The zero-risk theory of driver behavior and its implications. *Ergonomics*, *31*, 491–506.

Testable (2017). Testable (SaaS). Online: https://testable.org, accessed September 2017.

Treejack (2017). Optimal Workshop website (SaaS). Online: https://optimalworkshop.com/treejack, accessed September 2017.

Zhou, H., Itoh, M., Inagaki, T. (2008). Influence of cognitively distracting activity on driver's eye movement during preparation of changing lanes. In *Proceedings of the SICE Annual Conference 2008*, pp. 866–871.

11 Intuitive Interaction in Industry User Research
Context Is Everything

Jesyka M. Palmer, Temitope A. Ogunyoku,
and Christopher R. Hammond

CONTENTS

11.1 INTRODUCTION

This chapter examines intuitive interaction in the prototyping phase of a healthcare product designed for the emerging market of sub-Saharan Africa by a Western multinational. This research focused on commercial product development, which entails small sample sizes used iteratively over multiple phases of the development cycle. Differences were observed between participant groups, with the target population experiencing more anxiety and slower task completion rates than the non-target population. The chapter includes a proposal to better understand how a product's context of use affects test results and highlights the necessity for understanding local needs and experiences to conduct effective international research. Results are discussed and recommendations provided for Western multinationals creating offerings for users in emerging markets, who should take into account that intuitive interaction is contextual to the environment.

Stagnated US consumer markets is one factor causing Western multinationals to expand into emerging markets (Lauster, Mayr, Panneer, & Sehgal, 2010). Emerging markets are those countries with generally lower levels of household income and capital accumulation that have experienced brief economic growth and seen modest institutional changes (International Monetary Fund, 2017; MSCI Index Research, 2014). While different organizations—for instance, the International Monetary Fund, Morgan Stanley Capital International, Dow Jones, and MasterCard (Goldberg et al., 2008)—may each rank and classify countries differently on their emerging markets list, investors agree that emerging markets are attractive as they promise quick revenue and consumer growth for Western multinationals trying to meet their business objectives. A particular region of interest is sub-Saharan Africa, where recent research estimates that by 2020, middle-class consumers will grow to over 300 million individuals, GDP growth will average 5% a year, and consumer spending will reach $1 trillion (Adomaitis, Sychyk, & Tawii, 2016; Hatch, Becker, & van Zyl, 2011).

In the last few decades, many multinationals have attempted to break the barrier into emerging markets but have failed due to the lack of cultural understanding and inability to change their business positioning to fit local persons in the new environment—for example, Kellogg's failed attempt to introduce breakfast cereal into India after realizing that Indians prefer breakfast with warm foods, not cold milk ("A Humanist Who Knows Corn Flakes," 2005; Chavan, Gorney, Prabhu, & Arora, 2009). Walmart, a large one-stop shop for all consumer shopping needs, expanded into Latin American emerging markets and later withdrew after failing to adapt their business strategy, as Latin Americans prefer to shop around at smaller specialty shops (D'Andrea, Stengel, & Goebel-Krstelj, 2004). The Indian mobile phone company Bharti Airtel failed in their marketing campaign to South African individuals due to overgeneralization of the sub-Saharan African market (Manson, 2014). The common thread is Western and other non-local multinationals' failure to understand the user, and their needs, in their local environment.

The practice and application of intuitive interaction by Western multinationals continues to focus on ways to increase the efficiency of task interaction, improve the navigation in a user interface, or minimalize design (i.e., using less color, concise design elements, and simple typography). Intuitive interaction in the business environment is generally attached to the interface itself, as a product feature or benefit (Naumann et al., 2007). For practitioners, intuitive interaction can mean many things (see Chapters 1 and 2 of this book; Blackler, 2018; Blackler, Desai, McEwan, Popovic, & Diefenbach, 2018). But generally, in the practice of product development and design, intuitive interaction can be described as interaction that utilizes knowledge gained through prior experience (Blackler, Popovic & Mahar, 2003; Raskin, 1994), that requires no training, or in which the user is unaware of being trained (Spool, 2005). Due to the lack of shared prior experiences, it may be challenging for non-local multinational teams to create intuitive interactions within products that work for individuals in emerging markets.

This chapter examines a Western multinational's attempt to design a health-care product for use in Kenya. The first objective was to understand what analogous products already existed in the participants' environment and how this might affect user adoption. The second objective was to determine how to improve the product prior to users interacting with it based on usable interface design heuristics (Nielsen & Mack, 1994). Finally, the third objective was to determine how successfully users could interact with the product based on a series of tasks. After conducting qualitative and quantitative evaluation research, comparisons were drawn between how efficiently the target population and non-target population interacted with the product, including feedback on their experience. The chapter concludes with recommendations for those conducting evaluative research in emerging markets.

11.2 PRODUCT AND MATERIALS

The Cognitive Health Companion (CHC) kiosk was conceived by scientists at the IBM Research Africa (IBMRA) lab in Nairobi. The late-stage prototype CHC kiosk was undergoing product evaluation and was intended to assist in patient treatment of hypertension. This product is still under development and therefore IBM confidential; as such, an image cannot be included here. Non-communicable diseases (NCDs), which include hypertension, form a growing health burden within sub-Saharan Africa (Joshi et al., 2014; Olack et al., 2015; WHO, 2016). The CHC kiosk was a new self-service hypertension management product believed to be more efficient and accurate than existing products and services. It included commercially available hardware, integrated with an operating system running a series of applications that interacted, collected, analyzed, and continually delivered consumable health data to patients. The IBMRA engineering team built the CHC kiosk by connecting various devices to a retrofitted ultrasonic height and weight machine (UHWM) (model HGM-1 by Shengyuan, China). The attached devices that were tested included a Panasonic automatic arm blood pressure monitor (model EW3153 from Panasonic, China), an IriShield or USB iris scanner (model MK 2120U from Iritech, Fairfax, VA) used for user account security, and a tablet (Galaxy Tab S2 9.7 in. by Samsung, China) to facilitate communication with the user.

11.3 METHODOLOGY

The study (which ran from July 2016 to March 2017) was split into two phases. Phase 1 included environmental audits, which were conducted concurrently throughout the entire study. Phase 2 took place at IBMRA and included two evaluation methods: a heuristic evaluation followed by an in-lab usability test. All participants from both phases of the study were recruited from within IBMRA and screened to include members from both the target population (i.e., culturally Kenyan) and members from the non-target population (i.e., any other participant not from the target population). Each participant in phase 2 of the study signed a consent form prior to testing. Previous research shows that teams can find the majority of usability issues with five participants. Therefore, it is recommended that industry research keep sample sizes small to manage business and development resources, most importantly time and cost (Nielsen, 2000).

11.3.1   Environmental Audit

In phase 1, the team aimed to collect data about analogous products in context to understand potential barriers or opportunities to adoption and the accuracy of the health measurements. The product chosen to assess for this study was the Body Mass Index and Blood Pressure (BMIBP) kiosk. A review consisting of seven participants (four men and three women who were all from the target population) was conducted to determine whether the participants had ever seen a BMIBP kiosk and interacted with it, and their motivations for use. Results from this small survey were used to select three locations to conduct the environmental audits. The audits consisted of user observations and accuracy testing of three BMIBP kiosks: specifically, types of users, frequency of use, health measurement accuracy, and user experience.

11.3.2   Heuristic Evaluation

In phase 2, a heuristic evaluation, conducted by three user research professionals with backgrounds in product domain and usability, was applied to determine how to improve the CHC kiosk prior to users interacting with the product. Following Nielsen's 10 heuristics, or general usability principles, each researcher individually conducted an evaluation of the CHC kiosk twice: once without any assistance and a second time with an engineer-led demonstration (Nielsen, 1995a). Briefly, Nielsen's 10 heuristics are

1. Visibility of system status
2. Match between system and the real world
3. User control and freedom
4. Consistency and standards
5. Error prevention
6. Recognition rather than recall
7. Flexibility and efficiency of use
8. Aesthetic and minimal design
9. Help users recognize, diagnose, and recover from errors
10. Help and documentation

None of the researchers collaborated until they had documented their results. The three results were then combined into a single report prioritizing recommendations by severity level and type. These recommendations were implemented by the IBMRA engineering team and used to iterate the CHC kiosk prior to in-lab usability testing.

11.3.3   In-Lab Usability Testing

The aim of this method was to evaluate participants' perceived value of the CHC kiosk as well as the efficiency of their interactions. Participants included six individuals (four women, two men), split evenly among the user groups. Participants tested the CHC kiosk by attempting to complete the following subtasks:

1. Complete registration and iris scan
2. Record weight and height measurements
3. Complete two blood pressure measurements
4. Complete the short questionnaire
5. Understand the summary health risk score and the following steps

Participants were told that the purpose of the usability test was to evaluate the CHC kiosk and not to evaluate the participant. Each participant was instructed to approach the CHC kiosk and was given a scenario: *Imagine you are at a health-care facility and the health-care professional asks you to use the CHC kiosk to record your measurements.* Details about the nature of the CHC kiosk were added to this scenario in order to gather more constructive feedback as the testing progressed. Participants were instructed that at any point they wished to skip to the next subtask they could ask the moderator for help and move on. Once the participants believed they were finished with the subtasks, they could tell the moderator they were complete. The study utilized concurrent think-aloud protocol (Haak, Jong, & Schellens, 2003) as participants were asked by the moderators to communicate their thoughts and reactions to the CHC kiosk while performing the subtasks. Each usability test lasted about 30 minutes and was immediately followed by a post-usage online survey to record participant satisfaction across various attributes. These attributes were defined utilizing parts of Peter Morville's User Experience Honeycomb (2004): *useful, usable, desirable, findable, credible, accessible,* and *valuable.* The survey concluded with a standard Net Promoter Score (NPS) question: "How likely would you recommend the CHC to a friend or colleague?" The question is ranked on a scale from 0 to 10. Responses are categorized as Promoters (9–10), Passives (7–8), and Detractors (0–6) (Satmetrix Systems, Bain & Company, & Reichheld, 2017).

Qualitative and quantitative data collection for the performance and satisfaction-related measurements occurred via a video camera. The data were then analyzed by recording the participant's journey through each subtask into a spreadsheet to calculate the averages for both the target population and the non-target population (i.e., subtask completion time, subtask completion rate, and subtask error rate).

11.4 RESULTS

11.4.1 Environmental Audits

Although 71% ($N=5$) of participants remembered seeing BMIBP kiosks in a public setting, only 40% ($N=2$) of participants had used one. Reasons for not using BMIBP kiosks were that participants had other means of measuring their weight or were embarrassed because it publicly and verbally stated the user's results. The primary purpose of using the BMIBP kiosk was to track one's weight. None of the participants mentioned blood pressure tracking as a motivation for using a BMIBP kiosk. The majority of participants stated seeing BMIBP kiosks at similar places: at a specific mall in Nairobi and at three major streets in the Central Business District (CBD). However, participants mentioned that the BMIBP kiosks were not stationary and were moved by their owner to attract more business.

BMIBP kiosk 1 was located in a parking garage in between a parking pay station and a small street vendor that sold consumer retail items. No users used the kiosk during this observation. BMIBP kiosk 2 was located on Moi Ave in the CDB, next to *matatu* (i.e., public service vehicles) stations, restaurants, and stores. The location of the kiosk had high foot traffic but did not have any users during observation. BMIBP kiosk 3 was located near the Kenya Cinema, an environment with moderate foot traffic. Upon arrival, there was a user using the kiosk. The user only had his weight and height measured and the attendant provided little communication. Common traits across the BMIBP environmental kiosk audits were the following:

- Kiosks were in public settings with moderate to high volumes of foot traffic.
- An attendant was present and responsible for collecting payment and performing tasks to take users' measurements.
- Instructions (both images and written) as to how to use BMIBP kiosks were displayed on the front panel of the kiosk.
- Users were not encouraged and did not have time to read instructions because the attendants took measurements as soon as a user engaged with them.
- Interactions with the kiosk and attendant were less than 2 minutes.
- User health results were displayed on the screen and verbally announced by the kiosk.
- Users received a print-out of their results at the end of their interactions.

BMIBP kiosk health measurements were not accurate (Table 11.1). For all three measurements, the attendants did not state the need to remove shoes to obtain accurate results. The Kenya Hypertension Working Group guidelines state that a patient should wait quietly for 3–5 minutes before measuring their blood pressure, which was not observed in this audit. In addition, the blood pressure measurement should be done on both arms while the patient is sitting in a chair with back support and their arm supported at the level of their heart (Kenya Hypertension Working Group,

TABLE 11.1
BMIBP Kiosk Health Measurement Results

	Weight (kg)	Height (cm)	BMI	Blood Pressure (mmHg)	Pulse Rate (ppm)
BMIBP kiosk 1[a]	60.8	164.5	22.5	n/a	n/a
BMIBP kiosk 2[b]	56.5	164.5	20.9	106/76	83
BMIBP kiosk 3[b]	59	164.0	21.9	137/81	79
Average ± SD[c]	57.75 ± 1.77	164.25 ± 0.350	21.4 ± 0.710	127/78.5 ± 0.210	81 ± 2.83

[a] Kiosk used on July 7, 2016; attendant stated that the blood pressure machine was not working.
[b] Measurements taken on March 6, 2017.
[c] $N=2$ for measurements taken on March 6, 2017. SD = standard deviation.

2015). For all three BMIBP kiosks, the blood pressure was taken while standing on their feet and the blood pressure cuffs were not adjustable.

11.4.2 Heuristic Evaluation

As mentioned previously, the new CHC kiosk was a late-stage prototype and at times unstable and faulty, needing to be reset to complete subtasks. The intent was not to fix the hardware but to improve the interactions and interface of the CHC kiosk. No researcher successfully completed their evaluation without assistance from the IBMRA engineering team. Results of the heuristic evaluation were ranked based on severity level. Fifteen recommendations were provided, four of which were high severity or most likely to cause usability issues and lacked intuitive interactions. These four high-severity recommendations mapped to the following Nielsen heuristics: visibility of system status, user control and freedom, recognition rather than recall, and help users recognize, diagnose, and recover from errors (Nielsen, 1995b).

11.4.3 Usability Test

Performance-related data included subtask completion time, subtask completion rate, and subtask error rate. Completion time measured the total time for each participant to complete a subtask. Failure rate was a binary function that measured whether the participant was able to successfully complete the subtask without the moderator's assistance. Error rate was a count of the number of errors made by each participant while interacting with the CHC kiosk.

As shown in Table 11.2, on average the non-target population completed tasks at a rate that was 10.6% faster than the target population. The subtasks that contributed the most to this difference were subtasks 1 and 2, in which completion times almost doubled for the target population. The target population experienced a failure rate of 100% during subtask 1, while the failure rate of the non-target population was 66.7%. Only a single participant managed to successfully complete subtask 1 and that was by accident. No participant completed the overall tasks without assistance from the moderator. When considering error rate, the target population experienced roughly double the errors of the non-target population.

The average satisfaction scores were 9.33 and 5.00 for the target population and non-target population, respectively. The attributes with the greatest percentage difference between participant groups were useful (92.6%), desirable (85.7%), and usable (83%) (Table 11.3). The attribute with the least percentage difference between participant groups was *accessible* (11.8%). NPS scores were evenly split between Promoters and Detractors. All non-target population participants fell into the Detractor category. In the open-ended response to the NPS question, Detractors stated that they would like to see more personalized feedback from the CHC kiosk and easier input controls. All target population participants fell into the Promoter category. In the open-ended response, this group stated that the CHC kiosk was very easy to use, faster than other machines, and satisfactory.

The qualitative feedback from the usability test differed from the quantitative results. All non-target population participants interacted with the tablet integrated

TABLE 11.2
Usability Test Results (N = 6)

Subtask	1: Register and Record Iris Scan	2: Record Weight and Height	3: Record Two Blood Pressures	4: Complete Health Survey	5: Understand Health Risk Score
Target Population: P1, P5, P6[a]					
Completion time (seconds)	402 ± 126	176 ± 197	232 ± 70.5	61.0 ± 14.1	19.0 ± 20.8
Failure rate (%)	100	33.3	0	0	0
Error rate (count)	4.00 ± 1.00	0.667 ± 0.577	1.33 ± 0.577	0	0.667 ± 0.577
Non-target Population: P2, P3, P4[b]					
Completion time (s)	270 ± 82.8	70.3 ± 23.1	344 ± 258	51.7 ± 9.07	75.0 ± 12.7
Failure rate (%)	66.7	0	0	66.7	0
Error rate (count)	2.33 ± 1.15	0.333 ± 0.577	0.333 ± 0.577	0	0.500 ± 0.707

a Participant 6 could not complete subtask 4 and 5 due to machine malfunction.
b Participant 3 could not complete subtask 4 and 5 due to machine malfunction.

TABLE 11.3
Participant Satisfaction by Attribute (N = 6)

Satisfaction Scores	Overall Sat.[a]	Useful[b]	Usable	Desirable	Findable	Credible	Accessible	Valuable	NPS[c]
Avg. target population	9.33 ± 0.00	10.00 ± 0.00	9.67 ± 0.71	10.00 ± 0.00	9.67 ± 0.71	10.00 ± 0.00	9.00 ± 1.41	9.33 ± 1.41	10.00 ± 0.00
Avg. non-target population	5.00 ± 1.00	3.67 ± 2.08	4.00 ± 1.00	4.00 ± 1.00	4.33 ± 1.53	6.00 ± 2.65	8.00 ± 1.00	5.00 ± 0.00	2.00 ± 1.73

a Scale for overall satisfaction was 1 (extremely dissatisfied) to 10 (extremely satisfied).
b Scale for each attribute was 1 (extremely disagree) to 10 (extremely agree).
c NPS scale was 0 (extremely unlikely to recommend) to 10 (extremely likely to recommend).

with the CHC kiosk without asking the moderator for assistance. No participant in the target population group interacted with the tablet without first asking the moderator and assumed they did not need to interact with the CHC device for the service to initiate.

- P1 glanced around for a while and then stared at the tablet: "*I need to read?*"
- P5 stepped onto the scale and looked around without touching the tablet. After 141 seconds, P5 then asked, "*I should fill it in?*"
- P6 asked, "*Do I have to touch the machine? Do I talk to the machine or touch the button?*"

When continually prompted during tasks with think-aloud protocol, members of the target population continually waited for guidance from the moderator before performing activities, such as stepping onto the weight scale or using the blood pressure cuff. P6 said, "*I feel disappointed*" after completing the iris scan with help from the moderator. P1 stated, while waiting for weight and height measurements, "*I would be really anxious. I'm sick in the hospital and it's taking too long and I'm thinking I do not want to do this again.... It's taking forever.*"

The majority of non-target population participants focused on user interface improvements for the CHC kiosk to guide them through the subtasks. For example, P2 stated that they disliked the interaction with the tablet and the CHC kiosk: "*There is too much typing, too much reading; for me it would be better if it was in image form of how I was supposed to be, when to step on and when to step off.*" P3 expressed similar frustrations with the lack of feedback during their interactions: "*I'd like a bit more interactive feedback like 'inflating,' 'deflating' because it is all just the same message, even when I haven't placed my arm in the cuff.*" All participants in the non-target population group stated that there was too much reading involved. P4 admitted: "*I'm just scanning and impatient. I'm not reading it.*"

11.5 DISCUSSION

11.5.1 ADD FOUNDATIONAL USER RESEARCH PRIOR TO EVALUATION

Limited foundational research took place prior to the development of the CHC kiosk prototype. According to Spool (2005), this is common in Western multinationals and can lead to a large division in intuitive interaction design. Spool (2005) describes this division in a potential product design as the *Knowledge Gap* or the space between *Current Knowledge* and *Target Knowledge*. Current Knowledge represents the knowledge the user already has when they first see the interface. Target Knowledge is the knowledge the user actually needs to accomplish a given task (2005).

It was discovered in the environmental audits that the existing BMIBP kiosk was a quick catered service that required little input from the user. Although none of the target population participants had used a BMIBP kiosk, it is assumed that they were familiar with them due to the strategic placement of kiosks in high foot traffic areas; hence, this represented their Current Knowledge. Target Knowledge needed to complete tasks on the CHC kiosk consisted not only of the kiosk itself but also of the other integrated devices. While the IBMRA engineering team incorporated a familiar component (i.e., BMIBP kiosk) from the target population's daily environment, they did not fully consider the implication of attaching other less pervasive devices (e.g., tablet and iris scanner), which made the attendant from the Current Knowledge experience obsolete. Therefore, a large Knowledge Gap existed for target population participants interacting with the CHC kiosk. In fact, the majority of all participants failed subtask 1 (i.e., registration and iris scan). It is assumed the iris scan subtask was not an intuitive interaction. While the health results from the environmental audit suggest a more accurate product for collecting hypertensive measurements is needed in this market, there is a lack of user motivation. Reducing the Knowledge Gap requires development teams to conduct foundational research earlier to understand the target populations' current behaviors, associations, and specific needs.

11.5.2 Localize Heuristics and Consider Heuristic Evaluation User Tests

Results from the heuristic evaluation did not prepare the team for the usability test as intended. The evaluators of the CHC kiosk, being the Western multinational user researchers, were thought to suffice as evaluators because of their skills and expertise. Many of the usability problems captured by the evaluators were issues that resulted from slow keyboard interactions and inconsistent language and labels across the CHC kiosk experience.

Nevertheless, these issues were less severe for the target population for multiple reasons. For example, the research team originally rated "matching the system and the real world" (Nielsen, 1995b) at a low-to-medium severity level upon completion of the heuristic evaluation. Based on the results of the usability test, that rating should have been higher due to the cultural differences in intuitive interactions. Secondly, the research team focused on the user interface and did not take into account the physical and experiential attributes of the CHC kiosk. For example, BMIBP kiosks had attendants present to interact with the users, whereas the CHC kiosk was a self-service device. The BMIBP kiosks were faster than the CHC kiosk by 12 minutes on average. Finally, the BMIBP kiosks presented the complete health results at the end of the user's interaction, while the CHC kiosk iteratively presented results throughout the experience and lacked a take-home print-out. If these observed factors were considered prior to evaluation, the researchers may have been able to acknowledge them and use them to craft localized heuristics specific to the CHC kiosk's environment and target population.

Nielsen and Mack (1994) suggest using usability experts or users in a user test situation to conduct a heuristic evaluation. If the usability experts are not from the cultural background of the target population, it is essential that a heuristic evaluation be conducted in the user test situation with an experimenter to catch cultural biases of the usability experts. Weinschenk and Barker (2000) include a heuristic dedicated to meeting the cultural and social needs of the target population: *cultural propriety*. Although this heuristic exists, testing and conducting research with users from the target population is still needed to define what cultural propriety means in that context to ensure intuitive interactions are implemented into the product.

11.5.3   Adapt Usability Tests to Reduce Participant Bias

It is difficult to measure intuitive interactions within a product if the data being collected are biased. For example, the target population ranked the CHC kiosk high in terms of satisfaction despite the quite unpleasant and unsuccessful experience observed. Participant bias and demand characteristics from the target population may have been influenced by the Western multinational research team. This concurs with previous research where participants in India, when shown a technology they believed the foreign interviewer favored, were 2.5 times more likely to prefer that product even when an alternative was identical (Dell, Vaidyanathan, Medhi, Cutrell, & Thies, 2012). Other research shows that it is difficult for foreign interviewers to acquire constructive feedback as stranger interviews in Kenya produce less response validity and reliability (Bignami-Van Assche, Reniers, & Weinreb, 2003). Utilizing a professional from the local region to establish trust would be beneficial. as studies have suggested that participants responded more honestly to an interviewer's questions when the interviewer was from the same culture as the participant (Vatrapu & Perez-Quinones, 2006).

During the usability test, the protocol was iterated to gather more accurate feedback. Details were added to the original scenario description and the CHC kiosk introduction was changed from a neutral position to a negative position. Once the altered script claimed that the CHC kiosk was "broken" from the start, the target population began to provide more constructive feedback through the concurrent protocol. For instance, P1 and P5 both received the original script. P1 stated that the medical questionnaire used in the task provided good advice, even though she did not understand the poorly described medical terminology. P5 continually responded with the same phrase, "*It is interesting,*" when asked about her impressions. However, P6 received the changed script and provided responses such as "*Maybe the scanner is broken?*" and later contemplated: "*It's good but long. I'd put the blame on the machine.*"

More research needs to be done to understand why, after the script was changed, the survey responses did not change to provide more constructive feedback, while the qualitative feedback gathered through concurrent protocol did change for the target population. While adapting the moderator script did help provide more useful feedback from the concurrent think-aloud protocol, it did not help provide more constructive feedback from the survey responses. If only survey results were reported to the IBMRA engineering team, the results would suggest a different story about the target populations' satisfaction. Western multinationals conducting research in emerging markets should aim to collect both qualitative and quantitative data during evaluative studies to best understand the biases present and use these data to create intuitive products.

11.6   LIMITATIONS

The environmental audits were limited to three different locations as BMIBP kiosks were mobile and difficult to find. Due to the sensitive nature of conducting observations in the local environment, the research team was not able to stay in the environment for a long time period. The usability test was limited to six individuals, recruited using convenience sampling, and was confined to the lab as it was expensive to transport the CHC kiosk to a different location. However, based on the results of this study, it is recommended that Western multinationals conduct usability tests in the environment in which the product is intended to be located in order to better understand cultural differences that may emerge.

11.7 CONCLUSION

It is critical to understand the cultural context of a target population for the development of intuitive interaction within products. If there is a large cultural divide between the development team and target user, it is essential to conduct foundational research to understand the target population's unique needs and behaviors. One way for professionals to understand the user is to conduct field research. However, there are many challenges (e.g., participant bias, validity and reliability issues, etc.) associated with conducting evaluative studies in emerging markets (Dell et al., 2012; Bignami-Van Assche et al., 2006; Vatrapu & Perez-Quinones, 2006). Understanding the user and their needs can help craft heuristics to better interpret how well a product is serving that specific market. More heuristic adaption needs to occur in the practice of product development. Utilizing a mixed-method approach, by pairing quantitative methods with environmental audits and concurrent (think-aloud) protocol, can help capture cultural biases in usability tests. Western multinationals entering emerging markets need to continually incorporate the cultural context of the target population's need into their development cycle to create intuitive products that will be competitive in the marketplace and deliver high user satisfaction.

ACKNOWLEDGMENTS

IBMRA health-care team: Kala Fleming, Aisha Walcott, Reginald Bryant, Katherine Tryon, Meenal Pore, Charles Wachira, Oliver Bent, Simone Fobi Nsutezo, Daby Sow, and Erick Oduor.

REFERENCES

A Humanist Who Knows Corn Flakes. (2005). *Harvard Magazine*, September–October, 64–65.
Adomaitis, K., Sychyk, I., & Tawii, C. (2016). Doing business beyond South Africa: Growth and opportunity in sub-Saharan cities (Rep.). Retrieved April 25, 2017, from Euromonitor International website: http://go.euromonitor.com/rs/805-KOK-719/images/WP_Beyond-Capital-Cities_1.3-0716.pdf.
Bignami-Van Assche, S., Reniers, G., & Weinreb, A. A. (2003). An assessment of the KDICP and MDICP data quality: Interviewer effects, question reliability and sample attrition. *Demographic Research, Special 1*, 31–76.
Blackler, A. (2018). Intuitive interaction: An overview. In A. Blacker (Ed.), *Intuitive Interaction: Research and Application* (pp. 3–18). Boca Raton, FL: CRC Press.
Blackler, A., Desai, S., McEwan, M., Popovic, V., & Diefenbach, S. (2018). Perspectives on the nature of intuitive interaction. In A. Blackler (Ed.), *Intuitive Interaction: Research and Application* (pp. 19–40). Boca Raton, FL: CRC Press.

Blackler, A., Popovic, V., & Mahar, D. (2003). The nature of intuitive use of products: An experimental approach. *Design Studies*, *24*(6), 491–506.

Chavan, A. L., Gorney, D., Prabhu, B., & Arora, S. (2009). The washing machine that ate my sari: Mistakes in cross-cultural design. *Interactions* (January–February), 16(1): 26–31. doi:10.1145/1456202.1456209.

D'Andrea, G., Stengel, E. A., & Goebel-Krstelj, A. (2004). 6 Truths about emerging-market consumers. *Strategy Business*, *Spring*, 34, 2–12.

Dell, N., Vaidyanathan, V., Medhi, I., Cutrell, E., & Thies, W. (2012). Yours is better! *Proceedings of the 2012 ACM Annual Conference on Human Factors in Computing Systems (CHI '12)*, Austin, TX, pp. 1321–1330. doi:10.1145/2207676.2208589.

Goldberg, M., Hedrick-Wong, Y., Bhaskaran, M., Gang, F., Lever, W., Levi, M., Pellegrini, A., & Taylor, P. J. (2008). MasterCard Worldwide centers of commerce: Emerging markets index (Rep.). MasterCard Worldwide website. Retrieved April 25, 2017, from: http://www.mastercard.com/us/company/en/insights/pdfs/2008/MCWW_EMI-Report_2008.pdf.

Haak, M. V., Jong, M. D., & Schellens, P. J. (2003). Retrospective vs. concurrent think-aloud protocols: Testing the usability of an online library catalogue. *Behaviour & Information Technology*, *22*(5), 339–351.

Hatch, G., Becker, P., & Van Zyl, M. (2011). The dynamic African consumer market: Exploring growth opportunities in sub-Saharan Africa (Rep. No. 11-0457_LL / 7-1842). Accenture website. Retrieved April 25, 2017, from: http://www.finnpartnership.fi/__kehitysmaa-tieto__/241/Accenture-The-Dynamic-African-Consumer-Market-Exploring-Growth-Opportunities-in-Sub-Saharan-Africa.pdf.

International Monetary Fund. (2017). World Economic Outlook Database April 2017: WEO groups and aggregates information. Retrieved April 25, 2017, from https://www.imf.org/external/pubs/ft/weo/2017/01/weodata/groups.htm.

Lauster, S. M., Mayr, E., Panneer, G., & Sehgal, V. (2010). Roasted or fried: How to succeed with emerging market consumers (Rep.). Strategy& website, July 7. Retrieved from: https://www.strategyand.pwc.com/reports/roasted-fried-succeed-with-emerging, accessed April 25, 2017.

Manson, K. (2014). "One size fits all" marketing by global companies fails in Africa. *Financial Times*, March 16. Retrieved from https://www.ft.com/content/944c7018-a5f2-11e3-b9ed-00144feab7de, accessed April 25, 2017.

Joshi, M. D., Ayah, R., Njau, E. K., Wanjiru, R., Kayima, J. K., Njeru, E. K., & Mutai, K. K. (2014). Prevalence of hypertension and associated cardiovascular risk factors in an urban slum in Nairobi, Kenya: A population-based survey. *BMC Public Health*, *14*(1177), 1471–2458.

Kenya Hypertension Working Group. (2015). Protocol for the identification and management of hypertension in adults in primary care. May 19. Retrieved from http://www.jkuat.ac.ke/departments/hospital/wp-content/uploads/2015/08/Hypretension-Protocol.pdf, accessed May 4, 2017.

Morville, P. (2004). User experience design. Semantic Studios website, June 21. Retrieved from http://semanticstudios.com/user_experience_design, accessed April 25, 2017.

MSCI Index Research. (2014). MSCI market classification framework (publication). MSCI website. Retrieved April 25, 2017, from: https://www.msci.com/documents/1296102/1330218/MSCI_Market_Classification_Framework.pdf/d93e536f-cee1-4e12-9b69-ec3886ab8cc8.

Naumann, A., Hurtienne, J., Israel, J. H., Mohs, C., Kindsmüller, M. C., Meyer, H. A., & Hußlein, S. (2007). Intuitive use of user interfaces: Defining a vague concept. *Engineering Psychology and Cognitive Ergonomics*, pp. 128–136. Lecture Notes in Computer Science. doi:10.1007/978-3-540-73331-7_14.

Nielsen, J. (1995a). How to conduct a heuristic evaluation. Nielsen Norman Group website, January 1. Retrieved May 4, 2017, from https://www.nngroup.com/articles/how-to-conduct-a-heuristic-evaluation.

Nielsen, J. (1995b). 10 usability heuristics for user interface design. Nielsen Norman Group website, January 1. Retrieved, from https://www.nngroup.com/articles/ten-usability-heuristics, accessed May 4, 2017.

Nielsen, J. (2000). Why you only need to test with 5 users. Nielsen Norman Group website, March 19. Retrieved from https://www.nngroup.com/articles/why-you-only-need-to-test-with-5-users.

Nielsen, J., (1994). Heuristic evaluation. In Nielsen, J., & Mack, R. L. (Eds.) *Usability Inspection Methods*. New York: John Wiley, 25–62.

Olack, B., Wabwire-Mangen, F., Smeeth, L., Montgomery, J. M., Kiwanuka, N., Breiman, R. F. (2015). Risk factors of hypertension among adults aged 35–64 years living in an urban slum Nairobi, Kenya. *BMC Public Health*, *15*, 1251.

Raskin, J. (1994). Intuitive equals familiar. *Communications of the ACM*, *37*(9), 17.

Satmetrix Systems, Bain & Company, & Fred Reichheld. (2017). What is Net Promoter? Net Promoter website. Retrieved from https://www.netpromoter.com/know, accessed May 4, 2017.

Spool, J. (2005). What makes a design seem "intuitive"? Retrieved from https://articles.uie.com/design_intuitive.

Vatrapu, R., & Perez-Quinones, M. A. (2006). Culture and usability evaluation: The effects of culture in structured interviews. *Journal of Usability Studies*, *1*(4), 156–170.

Weinschenk, S., & Barker, D. T. (2000). *Designing Effective Speech Interfaces*. New York: John Wiley.

World Health Organization (WHO) (2016). UN Task Force: Kenya's fight against noncommunicable diseases aims to improve health, strengthen development. WHO website, October 6. Retrieved from: http://www.who.int/nmh/ncd-task-force/unf-kenya/en, accessed April 25, 2017.

Index